T0093619

Chance, Calculation and Life

Series Editor
Jean-Charles Pomerol

Chance, Calculation and Life

Edited by

Thierry Gaudin
Marie-Christine Maurel
Jean-Charles Pomerol

WILEY

First published 2021 in Great Britain and the United States by ISTE Ltd and John Wiley & Sons, Inc.

ISTE Ltd
27-37 St George's Road
London SW19 4EU
UK

www.iste.co.uk

John Wiley & Sons, Inc.
111 River Street
Hoboken, NJ 07030
USA

www.wiley.com

Library of Congress Control Number: 2020952689

British Library Cataloguing-in-Publication Data
A CIP record for this book is available from the British Library
ISBN 978-1-78630-667-8

Contents

Chapter 15. Chance, Contingency and the Origins of Life: Some Historical Issues 241
Antonio LAZCANO

Chapter 16. Chance, Complexity and the Idea of a Universal Ethics 249
Jean-Paul DELAHAYE

Preface

The *Centre Culturel International de Cerisy* proposes, each year from the end of May to early October and within the welcoming context of a 17th-century castle, a historic monument, meetings to bring together artists, researchers, teachers, students, social and economical actors, as well as the wider public interested in cultural and scientific exchanges.

A long cultural tradition

– Between 1910 and 1939, Paul Desjardins organized the famous "decades" in Pontigny abbey, to unite eminent personalities for debates on literary, social and political themes.

– In 1952, Anne Heurgon-Desjardins, while repairing the castle, created the *Centre Culturel* and continued, with her own personal touch, the work of her father.

– From 1977 to 2006, her daughters, Catherine Peyrou and Edith Heurgon, took the lead and brought a new dimension to the activities.

– Today, after the departure of Catherine and then of Jacques Peyrou, Cerisy continues under the management of Edith Heurgon and Dominique Peyrou, supported by Anne Peyrou-Bas and Christian Peyrou, also part of Cerisy castle's Civil Society, as well as with the support of an efficient and dedicated team led by Philippe Kister.

A like-minded original project

– They receive, in a prestigious setting far removed from urban disturbances and for a relatively long time period, people who are animated by the same attraction for discussion, in order to, through communal contemplation, invent new ideas and weave lasting connections.

– The Civil Society graciously puts the premises at the disposal of the *Association des Amis de Pontigny-Cerisy*, with no lucrative purpose and recognized for its public interest, currently presided over by Jean-Baptiste de Foucauld, the inspector general of finances.

A regularly supported action

– The *Centre Culturel*, the main means of action of the Association, has organized nearly *750 symposiums* broaching, through completely independent routes, the most diverse of themes. These symposiums have given rise, through various editors, to the publication of approximately *550 books*.

– The *Centre National du Livre* ensures continuous support to the organization and publication of the symposiums. The *territorial collectivities* (Normandy Region, department Council of the Manche, Coutances Mer and Bocage) and the *regional directorate of cultural affairs* bring their support to the Center, which also organizes, alongside the *Universities of Caen* and *Rennes 2*, encounters on themes concerning Normandy and the Great West.

– A *Cercle des Partenaires*, circle of partners, formed of enterprises, local collectives and public bodies, supports and even initiates *prospective* encounters on the main *contemporary challenges*.

– Since 2012, a new modern and accessible conference room has allowed for a new formula: the "Entretiens de la Laiterie", days for exchanges and debates, a collaborative effort from the partners of the Association.

Thierry GAUDIN
Marie-Christine MAUREL
Jean-Charles POMEROL
January 2021

Information:
CCIC, Le Château, 50210 Cerisy-la-Salle, France
Tel.: + 33 (0) 2 33 46 91 66
website: www.ccic-cerisy.asso.fr
email: info.cerisy@ccic-cerisy.asso.fr

Introduction

During the Cerisy week that we organized in 2016, on the theme of *Sciences de la vie, sciences de l'information* (Gaudin *et al.* 2018; *Life Sciences, Information Sciences*), many questions were raised about the role of chance in the development and evolution of living beings. With the aim of further investigating this very specific question, that is, the role of chance in life, we once again invited eminent specialists in life, evolution, physics, probability and philosophy to Cerisy, during the week of August 29 to September 5, 2019.

This book is a collation of the main contributions presented on this occasion. In the first part, "Randomness in all of its Aspects", we will try to identify the concept in its various avatars. This part begins with a presentation by Cristian S. Calude and Giuseppe Longo who distinguish, from a scientific and philosophical point of view, the three generally recognized types of chance: "Common, Biological and Quantum Randomness". We can define common chance as either: the meeting of two independent chains of causality; or, as an event for which we ignore the causality or for a law which we know is very sensitive to small disturbances, disturbances which as Calude and Longo discuss, may in fact exist below the sensitivity threshold of our measuring instruments. This sensitivity to small initial disturbances leads to chaos in the mathematical sense of the term, and was illustrated at Cerisy by the screening of Etienne Ghys's videos, as commented on by Grégory Miermont. For his part, Gilles Pagès presents the point of view of the probabilist, who does not acknowledge chance, only probabilities. He also reminds us of the importance of the Monte-Carlo method, which consists of the creation of randomness through computer simulations, and shows us that the process by which mathematician-computer scientists "draw random numbers" is not so simple. Pagès goes on to echo Paul Lévy's declaration: "Nature ignores chance". In a very enlightening

Introduction written by Thierry GAUDIN, Marie-Christine MAUREL and Jean-Charles POMEROL.

talk on the historical and etymological origins of words, Clarisse Herrenschmidt checks this statement by reminding us that many languages have made an allocation for the concept of chance and have found words to express it.

Niels Bohr's interpretation of quantum physics has continued to engage, even scientists, by introducing what seems to be an intrinsic probability. This helped to inspire an article by François Vannucci on wave-particle duality: quantum randomness but statistical determinism. Einstein never really accepted indeterminacy and we see that the challenge to "Bohrian" indeterminism continues; this was illustrated by a disturbing presentation by Stéphane Douady. This indeterminacy extends to astrophysics and the theory of multiple universes, as we will discover from the inspired pen of Michel Cassé.

Does chance govern our decisions? Mathias Pessiglione explains how and why the brain does not maximize an expected utility. Nonetheless, reflecting on this further, he explains the mechanism of random decision-making in the brain, thereby posing the question of free will. Chance as divine intervention, as has often been suggested, could be another name for the hand of God when he does not want to reveal himself. This naturally leads to philosophical and poetic reflections before moving on to the second part of this account, devoted to biology and evolution. We will therefore focus on the meaning of life as captured by the poetic intelligence of Georges Amar, on a meandering path from Hölderlin to William of Ockham and Spinoza, passing many poets, including Rimbaud, Baudelaire and many others. "*Le hasard et le divin*" (Chance and the Divine) is the apt title chosen by Bertrand Vergely to share with us his thoughts on randomness, necessity and inner life, culminating in a passage on chance and Grace. This first part will end with a semantic exploration of the words "chance" and "creation", as guided by Ivan Magrin-Chagnolleau who defines himself as an "artist-researcher". Beyond semantics, he invites us to reflect on the links between randomness and artistic creation. Randomness is the only source of creation in genetics, as per Jacques Monod's vision. However, forms may require subjectivity to exist in order for them to work. This brings us to the tipping point, or *kairos*, alluded to by Bertrand Vergely, a moment in which the artist and the researcher has but a split second in which to *grab it by the horns*.

In the second part, "Randomness, Biology and Evolution", we begin with the topical subject of epigenetics, through a brilliant chapter by David Sitbon and Jonathan B. Weitzmann. The latter very clearly defines the pathways of randomness in sexual reproduction, in terms of the pairing of alleles and mutations. They also explain the role of chromatin in gene expression, and how the environment is involved in this epigenetic process. In a well-documented article, Bernard Dujon explains how genes are exchanged between neighboring species, or more distantly, during a horizontal transfer. This process is well known in prokaryotes, in particular through the introduction of foreign DNA using retroviruses (a process used in gene therapy). In contrast, the extent of this phenomenon in multicellular eukaryotes is only just beginning to be understood. The acquisition of genes by horizontal transmission justifies the title: *"Quand l'acquis devient héréditaire"* (*When Acquisition becomes Hereditary*). The link between the environment and randomness becomes obvious when one considers that this horizontal acquisition results from encounters, of a random nature, within a given biotope (i.e. the role of environment).

From reproduction to evolution, it is but a step to join Philippe Grandcolas as he tries to show that evolution does not happen completely at random. Evolutionary pressure and the environment constrain evolution. Philippe Grandcolas discusses the significance of adaptive pressure and specific inheritance. By analyzing the notions of convergence and parallelism, as well as the presentation of some evolutionary experiments, Philippe Grandcolas qualifies Gould's examination on the eventual "replay" of evolution.

With Amaury Lambert, we remain in the field of evolution and focus, more precisely, on the transition from genotype to phenotype in terms of the constraints of Waddington's[1] epigenetic landscape. After distinguishing between the different types of randomness in evolution, Lambert models how the phenotype can resist genetic hazards, thereby providing a precise meaning for the notion of "canalization" as introduced by Waddington. From this standpoint, it becomes normal to wonder whether the emergence of life is the result of random chance. Asking himself this question, Antonio Lazcano gives us a remarkably well-sourced contribution on the history of scientific ideas pertaining to the origins of life, which leads to the conclusion, already glimpsed by Mayr[2], that biology cannot be reduced to physics and chemistry since "Nothing in biology makes sense, except in the light of evolution"[3]. Therefore, we can qualify biology as a "historical science". Like history, evolution has its constraints and its own contingency, and therefore, very probably, random chance played a part in the emergence of life. Establishing this, however, brings about the question, is evolution naturally accompanied by an increase in complexity and an increase in biodiversity?

This question is asked by Jean-Paul Delahaye. To try and give a precise meaning to this question, Jean-Paul Delahaye reminds us of the notions of Kolmogorov computational complexity and Benett structural complexity[4]. Like the Universe,

---

1 See the Waddington landscape images and references in Amaury Lambert's article.

2 See the reference in the article by Antonio Lazcano.

3 Quote from Théodore Dobjansky: *American Biology Teacher*, 35, 125–129, 1973.

4 See references in the article by Jean-Paul Delahaye.

which tends to become increasingly complex in the sense of Benett, Jean-Paul Delahaye hypothesizes that evolution is likewise a process that tends toward complexification, one that sometimes backtracks on itself with the result of it being erased. The growing complexity of living things is reflected upon in terms of biodiversity. There seems to be a certain human attraction to complexity, in particular by living organisms, a tropism that opens the door for a discussion on the ethics of organized complexity and it is along this line of questioning that Jean-Paul Delahaye, by way of recommendation, brings about his conclusion.

The purpose of this introduction is to give a brief overview on the richness of this work. From mathematics to the humanities, by way of biology, there are many questions and concepts linked to randomness to uncover. The contributions gathered here have the advantage of presenting the latest scientific results in a synthesized form, and with an abundant bibliography, that will serve both researchers and doctoral students. Yes, randomness is a part of life, evolution and reproduction, but always under certain constraints; importantly, according to the vision of quantum physics, individual randomness does not deter statistical determinism. This is a vast subject for multidisciplinary reflection that speaks to the irreducible individuation of man in a Universe that is increasingly explained and statistically deterministic.

Acknowledgments

This week could not have happened without the understanding of the many working colleagues and their laboratories, who carried some of the cost. We would like to thank EDF Île-de-France for the support given to the association of friends of Cerisy-Pontigny, which made it possible to organize this meeting. In addition, the CNRS institutes (INEE, INSB, INSMI, IN2P3) have also made their contribution. Finally, we would like to thank ISTE,who took over part of the editing, in particular, the English edition.

In memoriam

For two years, Dominique Lacroix helped us with the preparations for this conference, which blossomed from the one held in 2016. She unfortunately left us in June of 2019 and as such, was unable to meet with the speakers whom she knew. She would have been very happy to meet them again; however, the *chance* that is inherent to life decided otherwise. It is with great sadness that we dedicate this book to her.

Reference

Gaudin, T., Lacroix, D., Maurel, M.-C., Pomerol, J.-C. (eds) (2018). *Life Sciences, Information Sciences*. ISTE Ltd, London, and John Wiley & Sons, New York.

PART 1

Randomness in all of its Aspects

Classical, Quantum and Biological Randomness as Relative Unpredictability

We propose the thesis that randomness is unpredictability with respect to an intended theory and measurement. From this point of view, we briefly discuss various forms of randomness that physics, mathematics and computer science have proposed. Computer science allows us to discuss unpredictability in an abstract, yet very expressive way, which yields useful hierarchies of randomness and may help to relate its various forms in natural sciences. Finally, we discuss biological randomness – its peculiar nature and role in ontogenesis and in evolutionary dynamics (phylogenesis). Randomness in biology is positive as it contributes to organisms' and populations' structural stability by adaptation and diversity.

1.1. Introduction

Randomness is everywhere, for better or for worse: vagaries of weather, day-to-day fluctuations in the stock market, random motions of molecules or random genetic mutations are just a few examples. Random numbers have been used for more than 4,000 years, but they have never been in such high demand than they have in our time. What is the origin of randomness in nature and how does it relate to the only access we have to phenomena, that is, through measurement? How does randomness in nature relate to randomness in sequences of numbers? The theoretical and mathematical analysis of randomness is far from obvious. Moreover, as we will show, it depends on (and is relative to) the particular theory that is being worked on, the *intended* theoretical framework for the phenomena under investigation.

Chapter written by Cristian S. CALUDE and Giuseppe LONGO.

1.1.1. *Brief historical overview*

Democritus (460–370 BCE) determined the causes of things to necessity and chance alike, justifying, for example, the fact that atoms' disorderly motion can produce an orderly cosmos. However, the first philosopher to think about randomness was most likely Epicurus (341–270 BCE), who argued that "randomness is objective, it is the proper nature of events".

For centuries, though, randomness has only been mathematically analyzed in games and gambling. Luca Pacioli (in *Summa de aritmetica, geometria, proporzioni et proporzionalita*, 1494) studied how stakes had to be divided among gamblers, particularly in the difficult case when the game stops before the end. It is worth noting that Pacioli, a top Renaissance mathematician, also invented modern bookkeeping techniques (Double Entry): human activities, from gambling to financial investments, were considered as the locus for chance. As a matter of fact, early Renaissance Florence was the place of the invention of banks, paper currency and (risky) financial investments and loans[1].

Cardano (in *De Ludo Aleae* (The Game of Dice), 1525) developed Pacioli's analysis further. His book was only published in 1663, so Fermat and Pascal independently and more rigorously rediscovered the "laws of chance" for interrupted games in a famous exchange of letters in 1654. Pascal clarified the independence of history in the games of chance, against common sense: dice do not remember the previous drawings. Probabilities were generally considered as a tool for facing the lack of knowledge in human activities: in contrast to God, we cannot predict the future nor master the consequences of our (risky) actions. For the thinkers of the scientific revolution, randomness is not in nature, which is a perfect "Cartesian Mechanism": science is meant to discover the gears of its wonderful and exact mechanics. At most, as suggested by Spinoza, two independent, well-determined trajectories may meet (a walking man and a falling tile) and produce a random event. This may be considered a weak form of "epistemic" randomness, as the union of the two systems, if known, may yield a well-determined and predictable system and encounter.

Galileo, while still studying randomness, but only for dice (*Sopra le scoperte de i dadi*, 1612), was the first to relate measurement and probabilities (1632). For him, in physical measurement, errors are unavoidable, yet small errors are the most probable. Moreover, errors distribute symmetrically around the mean value, whose reliability increases with the number of measurements.

---

1 Such as the loan, in 1332, to the King of Britain Edward III who never returned it to the Bank of Bardi and Peruzzi – as all high school kids in Italy and our colleague Alberto Peruzzi in Florence know very well.

Almost two centuries later, Laplace brought Pascal's early insights to the modern rigor of probability theory (1998). He stressed the role of limited knowledge of phenomena in making predictions by equations: only a daemon with complete knowledge of all the forces in the Universe could "embrace in a single formula the movements of the greatest bodies of the universe and those of the tiniest atom; for such an intellect nothing would be uncertain and the future just like the past would be present before its eyes". The connection between incomplete knowledge of natural phenomena and randomness is made, yet no analysis of the possible "reasons" for randomness is proposed: probability theory gives a formal calculus of randomness, with no commitment on the nature of randomness. Defining randomness proved to be a hugely difficult problem which has only received acceptable answers in the last 100 years or so.

1.1.2. Preliminary remarks

Randomness is a tricky concept which can come in many flavors (Downey and Hirschfeldt 2010). Informally, randomness means *unpredictability, with a lack of patterns or correlations*. Why is randomness so difficult to understand and model? An intuitive understanding comes from the myriad of misconceptions and logical fallacies related to randomness, like the gambler's fallacy. In spite of the work of mathematicians since the Renaissance, there is the belief that after a coin has landed on tails 10 consecutive times, there are more chances that the coin will land on heads on the next flip. Similarly, common sense argues that there are "due" numbers in the lottery (since all numbers eventually appear, those that have not come up yet are "due", and thus more likely to come up soon). Each proposed definition of randomness seems to be doomed to be falsified by some more or less clever counter-example.

Even intuitively, the quality of randomness varies: tossing a coin may seem to produce a sequence of zeroes and ones which is less random than the randomness produced by Brownian motion. This is one of the reasons why users of randomness, like casinos, lotteries, polling firms, elections and clinical evaluations, are hard pressed to "prove" that their choices are "really" random. A new challenge is emerging, namely, to "prove randomness".

For physical systems, the randomness of a process needs to be differentiated from that of its outcome. Random (stochastic) processes have been extensively studied in probability theory, ergodic theory and information theory. Process or genesis randomness refers to such processes. On one hand, a "random" sequence does not necessarily need to be the output of a random process (e.g. a mathematically defined "random" sequence) and, conversely, a random process (e.g. a quantum random generator) is expected, but not guaranteed, to produce a "random output".

Outcome (or product) randomness provides a *prima facie* reason for the randomness of the process generating that outcome (Eagle 2005, p. 762), but, as argued in Frigg (2004, p. 431), process and outcome randomness are not extensionally equivalent. Process randomness has no mathematical formalization and can only be accessed/validated with theory or output randomness.

Measurement is a constant underlying issue: we may only associate a number with a "natural" process, by measurement. Most of the time we actually associate an interval (an approximation), an integer or a rational number as a form of counting or drawing.

Let us finally emphasize that, in spite of the existing theoretical differences in the understanding of randomness, our approach unifies the various forms of randomness in a relativized perspective:

> Randomness is unpredictability with respect to the intended theory and measurement.

We will move along this epistemological stand that will allow us to discuss and compare randomness in different theoretical contexts.

1.2. Randomness in classical dynamics

A major contribution to the contemporary understanding of randomness was given by Poincaré. By his "negative result" (his words) on the Three Body Problem (1892, relatively simple deterministic dynamics, see below), he proved that minor fluctuations or perturbations below the best possible measurement may manifest in a measurable, yet unpredictable consequence: "we then have a random phenomenon" (Poincaré 1902). This started the analysis of deterministic chaos, as his description of the phase-space trajectory derived from a nonlinear system is the first description of chaotic dynamics (Bros and Iagolnitzer 1973).

Poincaré's analysis is grounded in a mathematical "tour de force". He proved the non-analyticity of the (apparently) simple system of nonlinear equations describing two planets and a Sun in their gravitational fields (three bodies). The planets disturb each other's trajectories and this gives the formal divergence of the Lindstedt-Fourier series meant to give a linear approximation of the solution of the equations. More precisely, by using his notions of bifurcations and homoclinic orbit (the intersection of a stable and an unstable manifold), he showed that the "small divisors", which make the series diverge, physically mean that an undetectable fluctuation or perturbation may be amplified to a measurable quantity by the choice of a branch, or another in a bifurcation, or a manifold along a homoclinic orbit. It is often difficult to give a physical meaning to the solution of a system of equations; it is particularly hard and inventive to make sense of the absence of a solution. Yet, this made us understand randomness as deterministic unpredictability and non-analyticity as a strong form of classical unpredictability[2].

In this classical framework, a random event has a cause, yet this cause is below measurement. Thus, Curie's principle[3] is preserved: "the asymmetries of the consequences are already present in the causes" or "symmetries are preserved" – the asymmetries in the causes are just hidden.

For decades, Poincaré's approach was quoted and developed by only a few, that is, until Kolmogorov's work in the late 1950s and Lorentz in the 1960s. Turing is one of these few: he based his seminal paper on morphogenesis (Turing 1952) on the nonlinear dynamics of forms generated by chemical reactants. His "action/reaction/diffusion system" produced different forms by spontaneous symmetry breaking. An early hint of these ideas is given by him in Turing (1950, p. 440): "The displacement of a single electron by a billionth of a centimetre at one moment might make the difference between a man being killed by an avalanche a year later, or escaping". This Poincarian remark by Turing preceded by the famous "Lorentz butterfly effect" (proposed in 1972) by 20 years on the grounds of Lorentz's work from 1961.

Once more, many like to call this form of classical randomness "epistemic" unpredictability, that is related to our knowledge of the world. We do not deal with ontologies here, although this name may be fair, with the distinction from the understanding of randomness as a very weak form of unpredictability proposed by Spinoza. Poincaré brought a fact known since Galileo into the limelight: classical measurement is an interval, by principle[4]. Measurement is the only form of access we have to the physical world, while no principle forbids, *a priori*, to join two independent Spinozian dynamics. That is, even epistemic, classical physics posits

---

2 The non-analyticity is stronger than the presence of positive Lyapunov exponents for a nonlinear function. These exponents can appear in the solution of a nonlinear system or directly in a function describing a dynamic. They quantify how a minor difference in initial conditions can be magnified along a trajectory. In this case, we can have a form of "controlled" randomness because the divergence of the trajectories starting within the same best measurement interval will never exceed a pre-assumed, exponentially increasing value. In the absence of (analytical) solutions, bifurcations and homoclinic orbits can lead to sudden and "uncontrolled" divergence.

3 A macroscopic cause cannot have more elements of symmetry than the effects it produces. Its informational equivalent, called data processing inequality, asserts that no manipulation of information can improve the conclusions drawn from such data (Cover and Thomas 1991).

4 Laplace was also aware of this, but Lagrange, Laplace and Fourier firmly believed that any system of Cauchy equations possessed a linear approximation (Marinucci 2011).

this limit to access and knowledge as *a priori* measurement. Then this lack of complete knowledge yields classical randomness, typically in relation to a nonlinear mathematical modeling, which produces either positive Lyapunov exponents or, more strongly, non-analyticity. In other words, classical systems (as well as relativistic ones) are deterministic, yet they may be unpredictable, in the sense that randomness is not in the world nor it is just in the eyes of the beholder, but it pops out at the interface between us and the world by *theory* and *measurement*.

By "theory" we mean the equational or functional determination, possibly by a nonlinear system of equations or evolution functions.

1.3. Quantum randomness

Quantum randomness is hailed to be more than "epistemic", that is, "intrinsic" (to the theory). However, quantum randomness is not part of the standard mathematical model of the quantum which talks about probabilities, but is about the measurement of individual observables. So, to give more sense to the first statement we need to answer (at least) the following questions: (1) What is the source of quantum randomness? (2) What is the quality of quantum randomness? (3) Is quantum randomness different from classical randomness?

A naive answer to (1) is to say that quantum mechanics has shown "without doubt" that microscopic phenomena are intrinsically random. For example, we cannot predict with certainty how long it will take for a single unstable atom in a controlled environment to decay, even if one has complete knowledge of the "laws of physics" and the atom's initial conditions. One can only calculate the probability of decay in a given time, nothing more! This is intrinsic randomness guaranteed.

But is it? What is the cause of the above quantum mechanical effect? One way to answer is to consider a more fundamental quantum phenomenon: *quantum indeterminism*. What is quantum indeterminism and where does it come from? Quantum indeterminism appears in the measurement of individual observables: it has been at the heart of quantum mechanics since Born postulated that the modulus-squared of the wave function should be interpreted as a probability density that, unlike in classical statistical physics (Myrvold 2011), expresses fundamental, irreducible indeterminism (Born 1926). For example, the measurement of the spin, "up or down", of an electron, in the standard interpretation of the theory, is considered to be pure contingency, a symmetry breaking with no antecedent, in contrast to the causal understanding of Curie's principle[5]. The nature of individual measurement outcomes in quantum mechanics was, for a period, a subject of much debate. Einstein famously dissented, stating his belief that "He does not throw dice" (Born 1969, p. 204). Over time the assumption that measurement outcomes are fundamentally indeterministic became a postulate of the quantum orthodoxy (Zeilinger 2005). Of course, this view is not unanimously accepted (see Laloë 2012).

Following Einstein's approach (Einstein *et al.* 1935), quantum indeterminism corresponds to the absence of physical reality, if reality is what is made accessible by measurement: if no unique element of physical reality corresponding to a particular physical observable (thus, measurable) quantity exists, this is reflected by the physical quantity being indeterminate. This approach needs to be more precisely formalized. The notion of *value indefiniteness*, as it appears in the theorems of Bell (Bell 1966) and, particularly, Kochen and Specker (1967), has been used as a formal model of quantum indeterminism (Abbott *et al.* 2012). The model also has empirical support as these theorems have been experimentally tested via the violation of various inequalities (Weihs *et al.* 1998). We have to be aware that, going along this path, the "belief" in quantum indeterminism rests on the assumptions used by these theorems.

An observable is *value definite* for a given quantum system in a particular state if the measurement of that observable is pre-determined to take a (potentially hidden) value. If no such pre-determined value exists, the observable is value indefinite. Formally, this notion can be represented by a *(partial) value assignment function* (see Abbott *et al.* (2012) for the complete formalism).

When should we conclude that a physical quantity is value definite? Einstein, Podolsky and Rosen (EPR) defined *physical reality* in terms of certainty of predictability in Einstein *et al.* (1935, p. 777):

> If, without in any way disturbing a system, we can predict with certainty (i.e., with probability equal to unity) the value of a physical quantity, then there exists an element of reality corresponding to that quantity.

Note that both allusions to "disturbance" and to the (numerical) value of a physical quantity refer to measurement as the only form of access to reality we have. Thus, based on this accepted notion of an element of physical reality,

---

5 A correlation between random events and symmetry breakings is discussed in Longo *et al.* (2015). In this case, measurement produces a value (up or down), which breaks the in-determined or in-differentiated (thus, symmetric) situation before measurement.

following (Abbott *et al.* 2012) we answer the above question by identifying the EPR notion of an "element of physical reality" with "value definiteness":

EPR principle: If, without disturbing a system in any way, we can predict with certainty the value of a physical quantity, then there exists a *definite value* prior to the observation corresponding to this physical quantity.

The EPR principle justifies:

Eigenstate principle: a projection observable corresponding to the preparation basis of a quantum state is *value definite*.

The requirement called *admissibility* is used to avoid outcomes that are impossible to obtain according to quantum predictions, but which have overwhelming experimental confirmation:

Admissibility principle: definite values must not contradict the statistical quantum predictions for compatible observables of a single quantum.

Non-contextuality principle: the measurement results (when value definite) do not depend on any other compatible observable (i.e. simultaneously observable), which can be measured in parallel with the value definite observable.

The Kochen-Specker Theorem (Kochen and Specker 1967) states that no value assignment function can consistently make *all* observable values definite while maintaining the requirement that the values are assigned non-contextually. This is a global property: non-contextuality is incompatible with *all* observables being value definite. However, it is possible to localize value indefiniteness by proving that even the existence of two non-compatible value definite observables is in contradiction with admissibility and non-contextually, without requiring that all observables be value definite. As a consequence, we obtain the following "formal identification" of a value indefinite observable:

Any mismatch between preparation and measurement context leads to the measurement of a value indefinite observable.

This fact is stated formally in the following two theorems. As usual we denote the set of complex numbers by $\mathbb{C}$ and vectors in the Hilbert space $\mathbb{C}^n$ by $|.>$; the projection onto the linear subspace spanned by a non-zero vector $|\varphi>$ is denoted by P_φ. For more details see Laloë (2012).

THEOREM 1.1.– *Consider a quantum system prepared in the state |ψ> in dimension n ≥ 3 Hilbert space $\mathbb{C}^n$, and let |φ> in any state neither parallel nor orthogonal to |ψ>. Then the observable projection $P_φ$ is value indefinite under any non-contextual, admissible value assignment.*

Hence, accepting that definite values *exist* for certain observables (the eigenstate principle) and behave non-contextually (non-contextuality principle) is enough to locate and *derive, rather than postulate*, quantum value indefiniteness. In fact, *value indefinite observables* are far from being scarce (Abbott *et al.* 2014b).

THEOREM 1.2.– *Assume the eigenstate principle, non-contextuality and admissibility principles. Then, the (Lebesgue) probability that an arbitrary value indefinite observable is 1.*

Theorem 1.2 says that all value definite observables can be located in a small set of probability zero. Consequently, value definite observables are not the norm, they are the exception, a long time held intuition in quantum mechanics.

The above analysis not only offers an answer to question (1) from the beginning of this section, but also indicates a procedure to generate a form of quantum random bits (Calude and Svozil 2008; Abbott *et al.* 2012, 2014a): to *locate and measure a value indefinite observable*. Quantum random number generators based on Theorem 1.1 were proposed in (Abbott *et al.* 2012, 2014a). Of course, other possible sources of quantum randomness may be identified, so we are naturally led to question (2): what is the quality of quantum randomness certified by Theorem 1.1, and, if other forms of quantum randomness exist, what qualities do they have?

To this aim we are going to look, in more detail, at the unpredictability of quantum randomness certified by Theorem 1.1. We will start by describing a non-probabilistic model of prediction – proposed in (Abbott *et al.* 2015b) – for a hypothetical experiment E specified effectively by an experimenter[6].

The model uses the following key elements:

1) The specification of an experiment E for which the outcome must be predicted.

2) A predicting agent or "predictor", which must predict the outcome of the experiment.

---

6 The model does not assess the ability to make statistical predictions – as probabilistic models might – but rather the ability to predict precise measurement outcomes.

3) An extractor ξ is a physical device that the predictor uses to (uniformly) extract information pertinent to prediction that may be outside the scope of the experimental specification E. This could be, for example, the time, measurement of some parameter, iteration of the experiment, etc.

4) The uniform, algorithmic repetition of the experiment E.

In this model, a predictor is an effective (i.e. computational) method to uniformly predict the outcome of an experiment using finite information extracted (again, uniformly) from the experimental conditions along with the specification of the experiment, but independent from the results of the experiments. A predictor depends on an axiomatic, formalized theory, which allows the prediction to be made, i.e. to compute the "future". An experiment is predictable if any potential sequence of repetitions (of unbounded, but finite, length) can always be predicted correctly by such a predictor. To avoid prediction being successful just by chance, we require that the correct predictor – which can return a prediction or abstain (prediction withheld) – never makes a wrong prediction, no matter how many times it is required to make a new prediction (by repeating the experiment) and cannot abstain from making predictions indefinitely, i.e. the number of correct predictions can be made arbitrarily large by repeating the experiment enough times.

We consider a finitely specified physical experiment E producing a single bit $x \in \{0,1\}$. Such an experiment could, for example, be the measurement of a photon's polarization after it has passed through a 50:50 polarizing beam splitter, or simply the toss of a physical coin with initial conditions and experimental parameters specified finitely.

A particular trial of E is associated with the parameter λ, which fully describes the "state of the universe" in which the trial is run. This parameter is "an infinite quantity" – for example, an infinite sequence or a real number – structured in a way dependent on the intended theory. The result below, though, is independent of the theory. While λ is not in its entirety an obtainable quantity, it contains any information that may be pertinent to prediction. Any predictor can have practical access to a finite amount of this information. We can view a resource as one that can extract finite information, in order to predict the outcome of the experiment E.

An *extractor* is a physical device selecting a finite amount of information included in λ without altering the experiment E. It can be used by a predicting agent to examine the experiment and make predictions when the experiment is performed with parameter λ. So, the extractor produces a finite string of bits $\xi(\lambda)$. For example, $\xi(\lambda)$ may be an encoding of the result of the previous instantiation of E, or the time of day the experiment is performed.

A *predictor* for E is an algorithm (computable function) P_E which halts on every input and outputs either 0, 1 (cases in which P_E has made a prediction), or "prediction withheld". We interpret the last form of output as a refrain from making a prediction. The predictor P_E can utilize, as input, the information ξ (λ) selected by an extractor encoding relevant information for a particular instantiation of E, but must not disturb or interact with E in any way; that is, it must be *passive*.

A predictor P_E provides a *correct prediction* using the extractor ξ for an instantiation of E with parameter λ if, when taking as input ξ (λ), it outputs 0 or 1 (i.e. it does not refrain from making a prediction) and this output is equal to x, the result of the experiment.

Let us fix an extractor ξ. The predictor P_E is *k-correct for ξ* if there exists an $n \geq k$, such that when E is repeated n times with associated parameters $\lambda_1, ..., \lambda_n$ producing the outputs $x_1, x_2, ..., x_n$, P_E outputs the sequence (ξ (λ_1)), P_E (ξ (λ_2)), ..., P_E (ξ (λ_n)) with the following two properties:

1) no prediction in the sequence is incorrect, and

2) in the sequence, there are *k correct* predictions.

The repetition of E must follow an algorithmic procedure for resetting and repeating the experiment; generally, this will consist of a succession of events, with the procedure being "prepared, performed, the result (if any) recorded and E being reset".

The definition above captures the need to avoid correct predictions by chance by forcing more and more trials and predictions. If P_E is k-correct for ξ, then the probability that such a correct sequence would be produced by chance $\binom{n}{k} \times 3^{-n}$ is bounded by $\left(\frac{2}{3}\right)^k$; hence, it tends to zero when k goes to infinity.

The confidence we have in a k-correct predictor increases as k approaches infinity. If P_E is k-correct for ξ for all k, then P_E never makes an incorrect prediction and the number of correct predictions can be made arbitrarily large by repeating E enough times. In this case, we simply say that P_E is correct for ξ. The infinity used in the above definition is *potential*, not actual: its role is to arbitrarily guarantee many correct predictions.

This definition of correctness allows P_E to refrain from predicting when it is unable to. A predictor P_E which is correct for ξ is, when using the extracted information ξ (λ), guaranteed to always be capable of providing more correct predictions for E, so it will not output "prediction withheld" indefinitely. Furthermore, although P_E is technically only used a finite, but arbitrarily large, number of times, the definition guarantees that, in the hypothetical scenario where it is executed infinitely many times, P_E will provide infinitely many correct predictions and not a single incorrect one.

Finally, we define the prediction of a single bit produced by an individual trial of the experiment E. The outcome x of a single trial of the experiment E performed with parameter λ is predictable (with certainty) if there exist an extractor ξ and a predictor P_E which is correct for ⋯, and $P_E (\xi (\lambda)) = x$.

By applying the model of unpredictability described above to quantum measurement outcomes obtained by measuring a value indefinite observable, for example, obtained using Theorem 1.1, we obtain a formal certification of the unpredictability of those outcomes:

THEOREM 1.3. (Abbott *et al.* 2015b) – *Assume the EPR and Eigenstate principles. If E is an experiment measuring a quantum value indefinite observable, then for every predictor P_E using any extractor ξ, P_E is not correct for ξ.*

THEOREM 1.4. (Abbott *et al.* 2015b) – *Assume the EPR and Eigenstate principles. In an infinite independent repetition of an experiment E measuring a quantum value indefinite observable which generates an infinite sequence of outcomes $x = x_1x_2...$, no single bit x_i can be predicted with certainty.*

According to Theorems 1.3 and 1.4, the outcome of measuring a value indefinite observable is "maximally unpredictable". We can measure the degree of unpredictability using the computational power of the predictor.

In particular, we can consider weaker or stronger predictors than those used in Theorems 1.3 and 1.4, which have the power of a Turing machine (Abbott *et al.* 2015a). This "relativistic" understanding of unpredictability (fix the reference system and the invariant preserving transformations is the approach proposed by Einstein's relativity theory) allows us to obtain "maximal unpredictability", but not absolutely, only relative to a theory, no more and no less. In particular, and from this perspective, Theorem 1.3 should not be interpreted as a statement that quantum measurement outcomes are "true random"[7] in any absolute sense: true randomness – in the sense that no correlations exist between successive measurement results – is mathematically impossible as we will show in section 1.5 in a "theory invariant way", that is, for sequences of pure digits, independent of the measurements (classical, quantum, etc.) that they may have been derived from, if any. Finally, question (3) will be discussed in section 1.6.2.

---

7 Eagle argued that a physical process is random if it is "maximally unpredictable" (Eagle 2005).

1.4. Randomness in biology

Biological randomness is an even more complex issue. Both in phylogenesis and ontogenesis, randomness enhances variability and diversity; hence, it is core to biological dynamics. Each cell reproduction yields a (slightly) random distribution of proteomes[8], DNA and membrane changes, both largely due to random effects. In Longo and Montévil (2014a), this is described as a fundamental "critical transition", whose sensitivity to minor fluctuations, at transition, contributes to the formation of new coherence structures, within the cell and in its ecosystem. Typically, in a multicellular organism, the reconstruction of the cellular micro-environment, at cell doubling, from collagen to cell-to-cell connection and to the general tensegrity structure of the tissular matrix, all yield a changing coherence which contributes to variability and adaptation, from embryogenesis to aging. A similar phenomenon may be observed in an animal or plant ecosystem, a system yet to be described by a lesser coherence of the structure of correlations, in comparison to the global coherence of an organism[9].

Similarly, the irregularity in the morphogenesis of organs may be ascribed to randomness at the various levels concerned (cell reproduction and frictions/interactions in a tissue). Still, this is functional, as the irregularities of lung alveolus or of branching in vascular systems enhance ontogenetic adaptation (Fleury and Gordon 2012). Thus, we do not call these intrinsically random aspects of onto-phylogenesis "noise", but consider them as essential components of biological stability, a permanent production of diversity (Bravi Longo 2015). A population is stable because it is diverse "by low numbers": 1,000 individuals of an animal species in a valley are more stable if they are diverse. From low numbers in proteome splitting to populations, this contribution of randomness to stability is very different from stability derived from stochasticity in physics, typically in statistical physics, where it depends on huge numbers.

We next discuss a few different manifestations of randomness in biology and stress their positive role. Note that, as for the "nature" of randomness in biology, one must refer, at least, to both quantum and classical phenomena.

First, there exists massive evidence of the role of quantum random phenomena at the molecular level, with phenotypic effects (see Buiatti and Longo 2013 for an

---

8 Some molecular types are present in a few tenths or hundreds of molecules. Brownian motion may suffice to split them in slightly but non-irrelevantly different numbers.

9 An organism is an ecosystem inhabited by about 10^{14} bacteria, for example, and by an immune system, which in itself is an ecosystem (Flajnik and Kasahara 2010). Yet an ecosystem is not an organism: it has no relative metric stability (distance from its constituents), nor general organs of regulation and action, such as the nervous system found in animals.

introduction). A brand new discipline, quantum biology, studies applications of "non-trivial" quantum features such as superposition, non-locality, entanglement and tunneling to biological objects and problems (Ball 2011). "Tentative" examples include: (1) the absorbance of frequency-specific radiation, i.e. photosynthesis and vision; (2) the conversion of chemical energy into motion; and, (3) DNA mutation and activation of DNA transposons.

In principle, quantum coherence – a mathematical invariant for the wave function of each part of a system – would be destroyed almost instantly in the realm of a cell. Still, evidence of quantum coherence was found in the initial stage of photosynthesis (O'Reilly and Olaya-Castro 2014). Then, the problem remains: how can quantum coherence last long enough in a poorly controlled environment at ambient temperatures to be useful in photosynthesis? The issue is open, but it is possible that the organismal context (the cell) amplifies quantum phenomena by intracellular forms of "bio-resonance", a notion defined below.

Moreover, it has been shown that double proton transfer affects spontaneous mutation in RNA duplexes (Kwon and Zewail 2007). This suggests that the "indeterminism" in a mutation may also be given by *quantum randomness amplified by classical dynamics* (classical randomness, see section 1.6).

Thus, quantum events coexist with classical dynamics, including classical randomness, a hardly treated combination in physics – they should be understood in conjunction, for lack of a unified understanding of the respective fields. Finally, both forms of randomness contribute to the interference between levels of organization, due to regulation and integration, called bio-resonance in Buiatti and Longo (2013). Bio-resonance is part of the stabilization of organismal and ecosystemic structures, but may also be viewed as a form of proper biological randomness, when enhancing unpredictability. It corresponds to the interaction of different levels of organization, each possessing its own form of determination. Cell networks in tissues, organs as the result of morphogenetic processes, including intracellular organelles, are each given different forms of statistical or equational descriptions, mostly totally unrelated. However, an organ is made of tissues, and both levels interact during their genesis, as well as along their continual regeneration.

Second, for classical randomness, besides the cell-to-cell interactions within an organism (or among multicellular organisms in an ecosystem) or the various forms of bio-resonance (Buiatti and Longo 2013), let us focus on macromolecules' Brownian motion. As a key aspect of this approach, we observe that Brownian motion and related forms of random molecular paths and interactions must be given a fundamental and positive role in biology. This random activity corresponds to the thermic level in a cell, thus to a relevant component of the available energy: it turns out to be crucial for gene expression.

The functional role of stochastic effects has long since been known in enzyme induction (Novick and Weiner 1957), and even theorized for gene expression (Kupiec 1983). In the last decade (Elowitz *et al.* 2002), stochastic gene expression finally came into the limelight. The existing analyses are largely based on the classical Brownian motion (Arjun and van Oudenaarden 2008), while local quantum effects cannot be excluded.

> Increasingly, researchers have found that even genetically identical individuals can be very different, and that some of the most striking sources of this variability are random fluctuations in the expression of individual genes. Fundamentally, this is because the expression of a gene involves the discrete and inherently random biochemical reactions involved in the production of mRNA and proteins. The fact that DNA (and hence the genes encoded therein) is present in very low numbers means that these fluctuations do not just average away but can instead lead to easily detectable differences between otherwise identical cells; in other words, gene expression must be thought of as a stochastic process. (Arjun and van Oudenaarden 2008)

Different degrees of stochasticity in gene expression have been observed – with major differences in ranges of expression – in the same population (in the same organ or even tissue) (Chang *et al.* 2008).

A major consequence that we can derive from this view is the key role that we can attribute to this relevant component of the available energy, *heath*. The cell also uses it for gene expression instead of opposing to it. As a matter of fact, the view that DNA is a set of "instructions" (a program) proposes an understanding of the cascades from DNA to RNA to proteins in terms of a deterministic and predictable, thus programmable, sequence of stereospecific interactions (physico-chemical and geometric exact correspondences). That is, gene expression or genetic "information" is physically transmitted by these exact correspondences: stereospecificity is actually "necessary" for this (Monod 1970). The random movement of macromolecules is an obstacle that the "program" constantly fights. Indeed, both Shannon's transmission and Turing's elaboration of information, in spite of their theoretical differences, are both designed to oppose noise (see Longo *et al.* 2012b). Instead, in stochastic gene expression, Brownian motion, thus heath, is viewed as a positive contribution to the role of DNA.

Clearly, randomness, in a cell, an organism and an ecosystem, is highly constrained. The compartmentalization in a cell, the membrane, the tissue tensegrity structure, the integration and regulation by and within an organism, all contribute to restricting and canalizing randomness. Consider that an organism like ours has about 10^{13} cells, divided into many smaller parts, including nuclei: few physical structures are so compartmentalized. So, the very "sticky" oscillating and randomly moving macromolecules are forced within viable channels. Sometimes, though, it may not work, or it may work differently. This belongs to the exploration proper to biological dynamics: a "hopeful monster" (Dietrich 2003), if viable in a changing ecosystem, may yield a new possibility in ontogenesis, or even a new branch in evolution.

Activation of gene transcription, in these quasi-chaotic environments, with quasi-turbulent enthalpic oscillations of macro-molecules, is thus canalized by the cell structure, in particular in eukaryotic cells, and by the more or less restricted assembly of the protein complexes that initiate it (Kupiec 2010). In short, proteins can interact with multiple partners (they are "sticky") causing a great number of combinatorial possibilities. Yet, protein networks have a central hub where the connection density is the strongest and this peculiar canalization further forces statistical regularities (Bork *et al.* 2004). The various forms of canalization mentioned in the literature include some resulting from environmental constraints, which are increasingly acknowledged to produce "downwards" or Lamarckian inheritance or adaptation, mostly by a regulation of gene expression (Richards 2006). Even mutations may be both random and not random, highly constrained or even induced by the environment. For example, organismal activities, from tissular stresses to proteomic changes, can alter genomic sequences in response to environmental perturbations (Shapiro 2011).

By this role of constrained stochasticity in gene expression, molecular randomness in cells becomes a key source of the cell's activity. As we hinted above, Brownian motion, in particular, must be viewed as a positive component of the cell's dynamics (Munsky *et al.* 2009), instead of being considered as "noise" that opposes the elaboration of the genetic "program" or the transmission of genetic "information" by exact stereospecific macro-molecular interactions. Thus, this view radically departs from the understanding of the cell as a "Cartesian Mechanism" occasionally disturbed by noise (Monod 1970), as we give heath a constitutive, not a "disturbing" role (also for gene expression and not only for some molecular/enzymatic reactions).

The role of stochasticity in gene expression is increasingly accepted in genetics and may be generally summarized by saying that "macromolecular interactions, in a cell, are largely stochastic, they must be given in probabilities and the values of these probabilities depend on the context" (Longo and Montévil 2015). The DNA then becomes an immensely important physico-chemical trace of history, continually used by the cell and the organism. Its organization is stochastically used, but in a very canalized way, depending on the epigenetic context, to produce proteins and, at the organismal level, biological organization from a given cell. Random phenomena, Brownian random paths first, crucially contribute to this.

There is a third issue that is worth being mentioned. Following Gould (1997), we recall how the increasing phenotypic complexity along evolution (organisms become more "complex", if this notion is soundly defined) may be justified as a random complexification of the early bacteria along an asymmetric diffusion. The key point is to invent the right phase space for this analysis, as we hint: the tridimensional space of "biomass × complexity × time" (Longo and Montévil 2014b).

Note that the available energy consumption and transformation, thus entropy production, are the unavoidable physical processes underlying all biological activities, including reproduction with variation and motility, organisms' "default state" (Longo *et al.* 2015). Now, entropy production goes with energy dispersal, which is realized by random paths, as with any diffusion in physics.

At the origin of life, bacterial exponential proliferation was (relatively) unconstrained, as other forms of life did not contrast it. Increasing diversity, even in bacteria, by random differentiation started the early divergence of life, a process that would never stop – and a principle for Darwin. However, it also produced competition within a limited ecosystem and a slower exponential growth.

Gould (1989, 1997) uses the idea of random diversification to understand a blatant but often denied fact: life becomes increasingly "complex", if one accords a reasonable meaning to this notion. The increasing complexity of biological structures, whatever this may mean, has often been denied in order to oppose finalist and anthropocentric perspectives, where life is described as *aiming* at *Homo sapiens*, particularly at the reader of this chapter, the highest result of the (possibly intelligent) evolutionary path (or Design).

It is a fact that, under many reasonable measures, an eukaryotic cell is more complex than a bacterium; a metazoan, with its differentiated cells, tissues and organs, is more complex than a cell and that, by counting neurons and their connections, cell networks in mammals are more complex than in early triploblast (which have three tissues layers), and these have more complex networks of all sorts than diploblasts (like jellyfish, a very ancient life form). This nonlinear increase can be quantified by counting tissue differentiations, networks and more, as very informally suggested by Gould and more precisely quantified in (Bailly and Longo 2009) (see Longo and Montévil 2014b for a survey). The point is: how are we to understand this change toward complexity without invoking global aims?

Gould provides a remarkable, but very informal answer to this question. He bases it on the analysis of the asymmetric random diffusion of life, as constrained proliferation and diversification. Asymmetric because, by a common assumption, life cannot be less complex than bacterial life[10]. We may understand Gould's analysis by the general principle: *any asymmetric random diffusion propagates, by local interactions, the original symmetry breaking along the diffusion.*

The point is to propose a pertinent (phase) space for this diffusive phenomenon. For example, in a liquid, a drop of dye against a (left) wall diffuses in space (toward the right) when the particles bump against each other locally. That is, particles transitively inherit the original (left) wall asymmetry and propagate it *globally* by *local* random interactions. By considering the diffusion of *biomass*, after the early formation (and explosion) of life, over complexity, one can then apply the principle above to this fundamental evolutionary dynamic: biomass asymmetrically diffuses over complexity in time. Then, there is no need for a global design or aim: the random paths that compose *any* diffusion, also in this case, help to understand a random growth of complexity, on average. On average, as there may be local inversion in complexity, the asymmetry is randomly forced to "higher complexity", a notion to be defined formally, of course.

In Bailly and Longo (2009), and more informally in Longo and Montévil (2014b), a close definition of phenotypic complexity was given, by counting fractal dimensions, networks, tissue differentiations, etc., hence, a mathematical analysis of this phenomenon was developed. In short, in the suitable phase space, that is "biomass × complexity × time", we can give a diffusion equation with real coefficients, inspired by Schrödinger's equation (which is a diffusion equation, but in a Hilbert space). In a sense, while Schrödinger's equation is a *diffusion* of a law (an amplitude) of probability, the potential of variability of biomass over complexity in time was analyzed when the biological or phenotypic complexity was quantified, in a tentative but precise way, as hinted above (and better specified in the references).

Note that the idea that the complexity (however defined) of living organisms increases with time has been more recently adopted in Shanahan (2012) as a principle. It is thus assumed that there is a trend toward complexification and that this is intrinsic to evolution, while Darwin only assumed the divergence of characters. The very strong "principle" in Shanahan (2012), instead, may be *derived*, if one gives a due role to randomness, along an asymmetric diffusion, also in evolution.

---

10 Some may prefer to consider viruses as the least form of life. The issue is controversial, but it would not change Gould's and our perspective at all: we only need a minimum biological complexity which differs from inert matter.

A further but indirect fall-out of this approach to phenotypic complexity results from some recent collaborations with biologists of cancer (see Longo *et al.* 2015). We must first distinguish the notion of complexity, based on "counting" some key anatomical features, from biological organization. The first is given by the "anatomy" of a dead animal, and the second usually refers to the functional activities of a living organism. It seems that cancer is the only disease that diminishes functional organization by increasing complexity. When tissue is infected by cancer, ducts in glands, villi in epithelia, etc., increase in topological numbers (e.g. ducts have more lumina) and fractal dimensions (as for villi). This very growth of mathematical complexity reduces functionality, by reduced flow rates, thus the biological organization. This is probably a minor remark, but in the very obscure etiology of cancer, it may provide a hallmark for this devastating disease.

1.5. Random sequences: a theory invariant approach

Sequences are the simplest mathematical infinite objects. We use them to discuss some subtle differences in the quality of randomness. In evaluating the quality of randomness of the following four examples, we employ various tests of randomness for sequences, i.e. formal tests modeling properties or symptoms intuitively associated with randomness.

The *Champernowne sequence*, 012345678910111213, is random with respect to the statistical test which checks equal frequency – a clearly necessary condition of randomness[11]. Indeed, the digits 0, 1, 2, ..., 9 appear with the right frequency 10^{-1}, every string of two digits, like 23 or 00 appears with the frequency 10^{-2}, and by a classical result of Champernowne (1933), every string – say 366647888599991 00200030405060234234 or 000000000000000000000000000000000000 – appears with the frequency $10^{-(\text{length of string})}$ (10^{-35} in our examples). Is the condition sufficient to declare the Champernowne sequence random? Of course not. The Champernowne sequence is generated by a very simple algorithm – just concatenate all strings on the alphabet {0, 1, 2, 3, ..., 9} in increasing length order and use the lexicographical order for all strings of the same length. This algorithm allows for a prefect prediction of every element of this sequence, ruling out its randomness. A similar situation appears if we concatenate the prime numbers in base 10 obtaining the *Copeland-Erdös sequence* 235711131719232931374143 (Copeland and Erdös 1946).

---

11 This was the definition of chance informally proposed by Borel (1909): the proportion of times that a finite sequence is in an infinite (even very long) sequence corresponds (approximately) to the probability that such a sequence occurs during a particular test.

Now consider your favorite programing language L and note that each syntactically correct program has an end-marker (end or stop, for example) which makes correct programs self-delimited. We now define the binary *halting sequence* $H(L) = h_1 h_2 \ldots h_n$: enumerate all strings over the alphabet used by L in the same way as we did for the Champernowne sequence and define $h_i = 1$ if the ith string considered as a program stops and $h_i = 0$ otherwise. Most strings are not syntactically correct programs, so they will not halt: only some syntactically correct programs halt. The Church-Turing theorem – on the undecidability of the halting problem (Cooper 2004) – states that there is no algorithm which can correctly calculate (predict) all the bits of the sequence $H(L)$; so from the point of view of this randomness test, the sequence is random. Does $H(L)$ pass the frequency test? The answer is negative.

The Champernowne sequence and the halting sequence are both non-random, because each fails to pass a randomness test. However, each sequence passes a non-trivial random test. The test passed by the Champernowne sequence is "statistical", more quantitative, and the test passed by the halting sequence is more qualitative. Which sequence is "more random"?

Using the same programing language L we can define the *Omega sequence* as the binary expansion $\Omega(L) = \omega_1 \omega_2 \ldots \omega_n \ldots$ of the Omega number, the halting probability of L:

$$\sum_{halts} 2^{-(length\ of\ p)}$$

It has been proved that the Omega sequence passes both the frequency and the incomputability tests; hence, it is "more random" than the Champernowne, Copeland-Erdös and halting sequences (Calude 2002; Downey and Hirschfeldt 2010). In fact, the Omega sequence passes an infinity of distinct tests of randomness called Martin-Löf tests – technically making it Martin-Löf random (Calude 2002; Downey and Hirschfeldt 2010), one of the most robust and interesting forms of randomness.

Have we finally found the "true" definition of randomness? The answer is negative. A simple way to see it is via the following infinite set of computable correlations present in almost all sequences, including the Omega sequence (Calude and Staiger 2014), but not in all sequences: that is, for almost all infinite sequences, an integer $k > 1$ exists (depending on the sequence), such that for every $m \geq 1$:

$\omega_{m+1} \omega_{m+2} \ldots \omega_{mk} \neq 000 \ldots 0$ $m(k-1)$ times

In other words, every substring $\omega_{m+1} \omega_{m+2} \ldots \omega_{mk}$ has to contain at least one 1, for all $m \geq 1$, a "non-randomness" phenomenon no Martin-Löf test can detect. A more general result appears in Calude and Staiger (2014, Theorem 2.2).

So, the quest for a better definition continues! What about considering not just an incremental improvement over the previous definition, but a definition of "true randomness" or "perfect randomness"? If we are confined to just one intuitive meaning of randomness – the lack of correlations – the question becomes: Are there binary infinite sequences with no correlations? The answer is negative, so our quest is doomed to fail: there is no true randomness. We can prove this statement using the Ramsey[12] theory (see Graham and Spencer (1990) and Soifer (2011)) or the algorithmic information theory (Calude 2002).

Here is an illustration of the Ramsey-type argument. Let $s_1 \ldots s_n$ be a binary string. A *monochromatic* arithmetic progression of length k is a substring $s_i \, s_{i+t} \, s_{i+2t} \ldots s_{i+(k-1)t}$, $i \geq 1$ and $i + (k-1) \, t \leq n$ with all characters equal (0 or 1) for some $t > 0$. The string 01100110 contains no arithmetic progression of length 3 because the positions 1, 4, 5, 8 (for 0) and 2, 3, 6, 7 (for 1) do not contain an arithmetic progression of length 3; however, both strings 011001100 and 011001101 do: 1, 5, 9 for 0 and 3, 6, 9 for 1.

THEOREM 1.5.– Van der Waerden. *Every infinite binary sequence contains arbitrarily long monochromatic arithmetic progressions.*

This is one of the many results in Ramsey theory (Soifer 2011): it shows that in *any sequence* there are arbitrary long simple correlations. We note the power of this type of result: the property stated is true for *any sequence* (in contrast with typical results in probability theory where the property is true for *almost all sequences*). Graham and Spencer, well-known experts in this field, subtitled their *Scientific American* presentation of Ramsey Theory (Graham and Spencer 1990) with the following sentence:

> Complete disorder is an impossibility. Every large[13] set of numbers, points or objects necessarily contains a highly regular pattern.

Even if "true randomness" does not exist, can our intuition on randomness be cast in more rigorous terms? Randomness plays an essential role in probability theory, the mathematical calculus of random events. Kolmogorov axiomatic probability theory assigns probabilities to sets of outcomes and shows how to

---

12 The British mathematician and logician Frank P. Ramsey studied *conditions under which order must appear.*

13 The adjective "large" has precise definitions for both finite and infinite sets.

calculate with such probabilities; it assumes randomness, but does not distinguish between individually random and non-random elements. So, we are led to ask: is it possible to study classes of random sequences with precise "randomness/symptoms" of randomness? So far we have discussed two symptoms of randomness: statistical frequency and incomputability. More general symptoms are unpredictability (of which incomputability is a necessary but not sufficient form), incompressibility and typicality.

Algorithmic information theory (AIT) (Calude 2002; Downey and Hirschfeldt 2010), which was developed in the 1960s, provides a close analysis of randomness for sequences of numbers, either given by an abstract or a concrete (machine) computation, or produced by a series of physical measurements. By this, it provides a unique tool for a comparative analysis between different forms of randomness. AIT also shows that there is no infinite sequence passing all tests of randomness, so another proof that "true randomness" does not exist.

As we can guess from the discussion of the four sequences above, randomness can be refuted, but cannot be mathematically proved: we can never be sure that a sequence is "random", there are only forms and degrees of randomness (Calude 2002; Downey and Hirschfeldt 2010).

Finally, note that similarly to randomness in classical dynamics, which was made intelligible by Poincaré's negative result, AIT is also rooted in a negative result: Gödel's incompleteness theorem. As recalled above, a random sequence is a highly incomputable sequence. That is, algorithmic randomness is in a certain sense a refinement of Gödel's undecidability, as it gives a fine hierarchy of incomputable sets that may be related, as we will hint below, to relevant forms of randomness in natural sciences.

1.6. Classical and quantum randomness revisited

1.6.1. *Classical versus algorithmic randomness*

As we recalled in section 1.2, classical dynamical systems propose a form of randomness as unpredictability relative to a specific mathematical model and to the properties of measurement. Along these lines, Laskar (1994) recently gave an evaluation of the unpredictability of the position (and momentum) of planets in the Solar system, a system that has motivated all classical work since Newton and Laplace (their dynamics are unpredictable at relatively short astronomical time). How can one relate this form of deterministic unpredictability to algorithmic randomness, which is a form of pure mathematical incomputability? The first requires an interface between mathematical determination, by equations or evolution functions, and "physical reality" as accessed by measurement. The latter is a formal, asymptotic notion.

A mathematical relation may be established by considering Birkhoff ergodicity. This is a pure mathematical notion as it does not (explicitly) involve physical measurement, yet it applies to the nonlinear systems where one may refer to Poincaré's analysis or its variants, that is, to weaker forms of chaos based on positive Lyapunov exponents (see footnote 2) or sensitivity to initial conditions and "mixing" (see paragraph below). Birkhoff's notion is derived from his famous theorem and it roughly says that a trajectory, starting from a given point and with respect to a given observable quantity, is random when the value of a given observable over time coincides asymptotically with its value over space[14].

A survey of recent results that relate deterministic randomness – under the form of Birkhoff randomness for dynamical systems – to algorithmic randomness is presented in Longo (2012). This was mostly based on the relation between dynamical randomness and a weak form of Martin-Löf algorithmic randomness (due to Schnorr) (Galatolo *et al.* 2010; Gàcs *et al.* 2011). A subtle difference may be proved, by observing that some "typically" Birkhoff random points are not Martin-Löf random, some are even "pseudo-random", in the sense that they are actually computable. By restricting, in a physically sound way, the class of dynamical systems examined, a correspondence between points that satisfy the Birkhoff ergodic theorem and Martin-Löf randomness has been obtained in Franklin and Towsner (2014). These results require an "effectivization" of the spaces and dynamical theory, a non-obvious work, proper to all meaningful dynamical systems, as the language in which we talk about them is "effective": we formally write equations, solve them, when possible, and compute values, all by suitable algorithms. All of these results are asymptotic as they transfer the issue of the relation between theory and physical measurement to the limit behavior of a trajectory, as determined by the theory: a deterministic system sensitive to initial (or border) conditions, a mathematical notion, produces random infinite paths. To obtain these results it is enough to assume weak forms of deterministic chaos, such as the presence of dense trajectories, i.e. topological transitivity or mixing. The level of their sensitivity is reflected in the level of randomness of the trajectories, in the words of Franklin and Towsner (2014):

> Algorithmic randomness gives a precise way of characterizing how sensitive the ergodic theorem is to small changes in the underlying function.

---

14 Consider a gas particle and its momentum: the average value of the momentum over time (the time integral) is asymptotically assumed to coincide with the average *momenta* of all particles in the given, sufficiently large, volume (the space integral).

1.6.2. *Quantum versus algorithmic randomness*

We already summarized some of the connections between quantum and algorithmic unpredictability. The issue is increasingly in the limelight since there is a high demand for "random generators" in computing. Computers generate "random numbers" produced by algorithms and computer manufacturers took a long time to realize that randomness produced by software is only pseudo-random, that is, the generated sequences are perfectly computable, with no apparent regularity.

This form of randomness mimics the human perception of randomness well, but its quality is rather low because computability destroys many symptoms of randomness, e.g. unpredictability[15]. One of the reasons is that pseudo-random generators "silently fail over time, introducing biases that corrupt randomness" (Anthes 2011, p. 15).

Although, today, no computer or software manufacturer claims that their products can generate truly random numbers, these mathematically unfounded claims have re-appeared for randomness produced with physical experiments. They appear in papers published in prestigious journals, like Deutsch's famous paper (Deutsch 1985), which describes two quantum random generators (3.1) and (3.2) which produce "true randomness" or the *Nature 2010* editorial (titled *True Randomness Demonstrated* (Pironio *et al.* 2010)). Companies market "true random bits" which are produced by a "True Random Number Generator Exploiting Quantum Physics" (ID Quantique) or a "True Random Number Generator" (MAGIQ). "True randomness" does not necessarily come from the quantum. For example, "RANDOM.ORG offers true random numbers to anyone on the Internet" (Random.Org) using atmospheric noise.

Evaluating the quality of quantum randomness can now be done in a more precise framework. In particular, we can answer the question: is a sequence produced by repeated outcomes of measurements of a value indefinite observable computable? The answer is negative, in a strong sense (Calude and Svozil 2008; Abbott *et al.* 2012):

THEOREM 1.6.– *The sequence obtained by indefinitely repeating the measurement of a value indefinite observable, under the conditions of Theorem 1.1, produces a*

---

15 It is not unreasonable to hypothesize that pseudo-randomness reflects its creators' subjective "understanding" and "projection" of randomness. Psychologists have known for a long time that people tend to distrust streaks in a series of random bits; hence, they imagine a coin flipping sequence alternates between heads and tails much too often for its own sake of "randomness". As we said, the gambler's fallacy is an example.

bi-immune sequence (a strong form of incomputable sequence for which any algorithm can compute only finitely many exact bits).

Incomputability appears *maximally* in two forms:

– *Individualized*: no single bit can be predicted with certainty (Theorem 1.4), i.e. an algorithmic computation of a single bit, even if correct, cannot be formally certified;

– *asymptotic* (Theorem 1.6): only finitely many bits can be correctly predicted via an algorithmic computation.

It is an open question whether a sequence (such as Theorem 1.6) is Martin-Löf random or Schnorr random (Calude 2002; Downey and Hirschfeldt 2010).

1.7. Conclusion and opening: toward a proper biological randomness

The relevance of randomness in mathematics and in natural sciences further vindicates Poincaré's views against Hilbert's. The first has stressed (since 1890) his interest in negative results, such as his Three Body Theorem, and further claimed in *Science et Méthode* (1908) that "... unsolvable problems have become the most interesting and raised further problems to which we could not think before".

This is in sharp contrast with Hilbert's credo – motivated by his 1900 conjecture on the formal provability (decidability) of consistency of arithmetic – presented in his address to the 1930 Königsberg Conference[16] (Hilbert 1930):

For the mathematician, there is no unsolvable problem. In contrast to the foolish *Ignorabimus*, our credo is: We must know, We shall know.

As a matter of fact, the investigation of theoretical *ignorabimus*, e.g. the analysis of randomness, opened the way to a new type of knowledge, which does not need to give yes or no answers, but raises new questions and proposes new perspectives and, possibly, answers – based on the role of randomness, for example, a demonstrable mathematical unpredictability from which the geometry of Poincaré's dynamic systems is derived, or even the subsequent notions of an attractor, etc. Hilbert was certainly aware of Poincaré's unpredictability, but (or thus?) he limited his conjectures of completeness and decidability to formal systems, i.e. to *pure mathematical statements*. Poincaré's result instead, as recalled above, makes sense at the interface of mathematics and the physical world; for an analysis of the

---

16 Incidentally, the same conference at which Gödel presented his famous *incompleteness* theorem.

methodological links with Gödel's theorem, purely mathematical, and Einstein's result on the alleged incompleteness of quantum mechanics, see Longo (2018).

The results mentioned in section 1.6.1 turn, by using Birkhoff randomness in dynamical systems, physico-mathematical unpredictability into a pure mathematical form: they give relevant relations between the two frames, by embedding the *formal theory* of some physical dynamics into a computational frame, in order to analyze unpredictability with respect to that theory.

As for biology, the analysis of randomness at all levels of biological organization, from molecular activities to organismal interactions, clearly plays an increasing role in contemporary work. Still, we are far from a sound unified frame. First, because of the lack of unity even in the fragments of advanced physical analysis in molecular biology (typically, the superposition of classical and quantum randomness in a cell), to which one should add the hydrodynamical effect and the so-called "coherence" of water in cells pioneered by del Giudice (2007).

Second, biology has to face another fundamental problem. If one assumes a Darwinian perspective and considers phenotypes and organisms as proper biological observables, then the evolutionary dynamics imply a change to the very space of (parameters and) observables. That is, a phylogenetic analysis cannot be based on the *a priori* physical knowledge, the so-called condition of possibility for physico-mathematical theories: space-time and the pertinent observable, i.e. the phase space. In other words, a phylogenetic path cannot be analyzed in a pre-given phase space, like in all physical theories, including quantum mechanics, where self-adjoint operators in a Hilbert space describe observables. Evolution, by the complex blend of organisms and their ecosystem, co-constitutes its own phase space and this in a (highly) unpredictable way. Thus, random events, in biology, do not just "modify" the numerical values of an observable in a pre-given phase space, like in physics: they modify the very biological observables, the phenotypes, which is more closely argued in Longo and Montévil (2014b). If we are right, this poses a major challenge. In the analysis of evolutionary dynamics, randomness may not be measurable by probabilities (Longo *et al.* 2012a; Longo and Montévil 2014a). This departs from the many centuries of discussion on chance only expressed by probabilities.

If the reader were observing Burgess fauna, some 520 million years ago (Gould 1989), they would not be able to attribute probabilities to the changes of survival of Anomalocaris or Hallucigenia, or to one of the little chordates, nor their probabilities to become a squid, a bivalve or a kangaroo. To the challenges of synchronic measurement, proper to physical state determined systems, such as the ones we examined above, one has to add the even harder approximation of diachronic measurement, in view of the relevance of history in the determination of the biological state of affairs (Longo 2017).

Note that in section 1.4 we could have made a synthetic prediction: phenotypic complexity increases along evolution by a random, but asymmetric, diffusion. This conclusion would have been based on a global evaluation, a sum of all the numbers we associated with biological forms (fractal dimensions, networks' numbers, tissue folding ... all summed up). In no way, however, can you "project" this global value into specific phenotypes. There is no way to know if the increasing complexity could be due to the transformation of the lungs of early tetrapods into swim bladders and gills (branchia), or of their "twin-jointed jaw" into the mammalian auditory ossicles (Luo 2011).

Can AIT be of any help to meet this challenge? From a highly epistemic perspective, one may say that the phenotypes cannot be described before they appear. Thus, they form an incompressible list of described/describable phenotypes, at each moment of the evolutionary process. It seems hard to turn such a contingent/linguistic statement into an objective scientific analysis.

In contrast to AIT, in which randomness is developed for infinite sequences of numbers, in a measurement independent way, any study of randomness in a specific context, physical or biological, for example, depends on the *intended theory* which includes its *main assumptions* (see, for example, section 1.3 for the "principles" used in the analysis of quantum randomness and for the theory-and-measurement-dependent notion of the predictor).

As a consequence, our focus on unpredictability and randomness in natural sciences, where access to knowledge crucially requires physical or biological measurement, cannot and should not be interpreted as an argument that "the world is random" and even less that it is "computable" – we, historical and linguistic humans, effectively write our theories and compute them in order to predict (see Calude *et al.* 2012 for a discussion on the hypothesis of a lawless (physical) Universe).

We used (strong forms of) relative unpredictability as a tool to compare different forms of determination and stability in natural sciences, sometimes by a conceptual or mathematical duality (showing the "relevance of negative results"), other times by stressing the role of randomness in the robustness or resilience of phenomena. The ways we acquire knowledge may be enlightened by this approach, also in view of the role of symmetries and invariance in mathematical modeling and of the strong connections between spontaneous symmetry breaking and random events discussed in Longo and Montévil (2015).

1.8. Acknowledgments

The authors have been supported in part by Marie Curie FP7-PEOPLE-2010-IRSES Grant and have benefitted from discussions and collaboration with A. Abbott, S. Galatolo, M. Hoyrup, T. Paul and K. Svozil. We also thank the referees for their excellent comments and suggestions. We also warmly thank Springer Editions for authorizing the translation of the original article (Classical, Quantum and Biological Randomness as Relative Unpredictability. *Invited Paper*, special issue of *Natural Computing*, Volume 15, Issue 2, pp. 263–278, Springer, March 2016) and Louis Ter Ovanessian who provided the translation.

1.9. References

Abbott, A.A., Calude, C.S., Conder, J., Svozil, K. (2012). Strong Kochen-Specker theorem and incomputability of quantum randomness. *Physical Review A*, 86, 062109 [Online]. Available at: http://dx.doi.org/10.1103/PhysRevA.86.062109 [Accessed January 2021].

Abbott, A.A., Calude, C.S., Svozil, K. (2014a). Value-indefinite observables are almost everywhere. *Physical Review A*, 89, 032109 [Online]. Available at: http://dx.doi.org/10.1103/PhysRevA.89.032109 [Accessed January 2021].

Abbott, A.A., Calude, C.S., Svozil, K. (2014b). Value indefiniteness is almost everywhere. *Physical Review A*, 89(3), 032109 [Online]. Available at: http://arxiv.org/abs/1309.7188 [Accessed January 2021].

Abbott, A.A., Calude, C.S., Svozil, K. (2015a). A non-probabilistic model of relativised predictability in physics. *Information*, 6, 773–789.

Abbott, A.A., Calude, C.S., Svozil, K. (2015b). On the unpredictability of individual quantum measurement outcomes. In *Fields of Logic and Computation II – Essays Dedicated to Yuri Gurevich on the Occasion of His 75th Birthday, Lecture Notes in Computer Science*, Beklemishev, L.D., Blass, A., Dershowitz, N., Finkbeiner, B., Schulte, W. (eds). Springer, 9300, 69–86 [Online]. Available at: http://dx.doi.org/10.1007/978-3-319-23534-9_4 [Accessed January 2021].

Anthes, G. (2011). The quest for randomness. *Communications of the ACM*, 54(4), 13–15.

Arjun, R. and van Oudenaarden, R. (2008). Stochastic gene expression and its consequences. *Cell*, 135(2), 216–226.

Bailly, F. and Longo, G. (2009). Biological organization and anti-entropy. *Journal of Biological Systems*, 17(1), 63–96.

Ball, P. (2011). The dawn of quantum biology. *Nature*, 474, 272–274.

Barbara Bravi, G.L. (2015). The unconventionality of nature: Biology, from noise to functional randomness. In *Unconventional Computation and Natural Computation Conference*, Calude, C.S., Dinneen, M.J. (eds). Springer, LNCS 9252 [Online]. Available at: http://www.di.ens.fr/users/longo/files/CIM/Unconventional-NatureUCNC2015.pdf [Accessed January 2021].

Bell, J.S. (1966). On the problem of hidden variables in quantum mechanics. *Reviews of Modern Physics*, 38, 447–452 [Online]. Available at: http://dx.doi.org/10.1103/RevModPhys.38.447.

Borel, E. (1909). Les probabilités dénombrables et leurs applications arithmétiques. *Rendiconti del Circolo Matematico di Palermo (1884–1940)*, 27, 247–271 [Online]. Available at: http://dx.doi.org/10.1007/BF03019651 [Accessed January 2021].

Bork, P., Jensen, L.J., von Mering, C., Ramani, A.K., Lee, I., Marcotte, E.M. (2004). Protein interaction networks from yeast tohuman. *Current Opinion in Structural Biology*, 14, 292–299.

Born, M. (1926). Zur Quantenmechanik der Stoßvorgänge. *Zeitschrift für Physik*, 37, 863–867 [Online]. Available at: http://dx.doi.org/10.1007/BF01397477 [Accessed January 2021].

Born, M. (1969). *Physics in my Generation*, 2nd edition. Springer, New York.

Bros, J. and Iagolnitzer, D. (1973). Causality and local mathematical analyticity: Study. *Ann. Inst. Henri Poincaré*, 18(2), 147–184.

Buiatti, M. (2003). Functional dynamics of living systems and genetic engineering. *Rivista di biologia*, 97(3), 379–408.

Buiatti, M. and Longo, G. (2013). Randomness and multilevel interactions in biology. *Theory Bioscience*, 132, 139–158.

Calude, C. (2002). *Information and Randomness – An Algorithmic Perspective*, 2nd edition. Springer, Berlin.

Calude, C.S. and Staiger, L. (2018). Liouville numbers, Borel normality and algorithmic randomness. *Theory of Computing Systems*, 62(7), 1573–1585.

Calude, C.S. and Svozil, K. (2008). Quantum randomness and value indefiniteness. *Advanced Science Letters*, 1(2), 165–168 [Online]. Available at: http://dx.doi.org/10.1166/asl.2008.016.EprintarXiv:quant-ph/0611029 [Accessed January 2021].

Calude, C.S., Meyerstein, W., Salomaa, A. (2012). The universe is lawless or "pantôn chrêmatôn metron anthrôpon einai". *A Computable Universe: Understanding Computation & Exploring Nature as Computation*, Zenil, H. (ed.). World Scientific, Singapore.

Champernowne, D.G. (1933). The construction of decimals normal in the scale of ten. *The Journal of London Mathematical Society*, 8, 254–260.

Chang, H.H., Hemberg, M., Barahona, M., Ingber, D.E., Huang, S. (2008). Transcription wide noise control lineage choice in mammalian progenitor cells. *Nature*, 453, 544–548.

32   Chance, Calculation and Life

Cooper, S.B. (2004). *Computability Theory*. Chapman Hall/CRC Mathematics Series, Boca Raton.

Copeland, A.H. and Erdös, P. (1946). Note on normal numbers. *Bull. Amer. Math. Soc.*, 52, 857–860.

Cover, T.M. and Thomas, J.A. (1991). *Elements of Information Theory*. John Wiley & Sons, New York.

Deutsch, D. (1985). Quantum theory, the Church-Turing principle and the universal quantum computer. *Proceedings of the Royal Society of London. Series A, Mathematical and Physical Sciences (1934–1990)*, 400(1818), 97–117 [Online]. Available at: http://dx.doi.org/10.1098/rspa.1985.0070 [Accessed January 2021].

Dietrich, M. (2003). Richard Goldschmidt: Hopeful monsters and other "heresies". *Nature Reviews Genetics: Historical Article Journal Article Portraits*, 4, 68–74.

Downey, R. and Hirschfeldt, D. (2010). *Algorithmic Randomness and Complexity*. Springer, Berlin.

Eagle, A. (2005). Randomness is unpredictability. *British Journal for the Philosophy of Science*, 56(4), 749–790 [Accessed January 2021].

Einstein, A., Podolsky, B., Rosen, N. (1935). Can quantum-mechanical description of physical reality be considered complete? *Physical Review*, 47(10), 777–780 [Online]. Available at: http://dx.doi.org/10.1103/PhysRev.47.777 [Accessed January 2021].

Elowitz, M.B., Levine, A.J., Siggia, E.D., Swain, P.S. (2002). Stochastic gene expression in a single cell. *Science*, 297(5584), 1183–1186 [Online]. Available at: http://www.sciencemag.org/cgi/content/abstract/297/5584/1183 [Accessed January 2021].

Flajnik, M.F. and Kasahara, M. (2010). Origin and evolution of the adaptive immune system: Genetic events and selective pressures. *NatRev. Genet.*, 11(1), 47–59 [Online]. Available at: http://dx.doi.org/10.1038/nrg2703 [Accessed January 2021].

Fleury, V. and Gordon, R. (2012). Coupling of growth, differentiation and morphogenesis: An integrated approach to design in embryogenesis. In *Origin(s) of Design in Nature, Cellular Origin, Life in Extreme Habitats and Astrobiology*, Volume 23, Swan, L., Gordon, R., Seckbach, J. (eds). Springer, Netherlands [Online]. Available at: http://dx.doi.org/10.1007/978-94-007-4156-0_22 [Accessed January 2021].

Franklin, J.N. and Towsner, H. (2014). *Randomness and Non-ergodic Systems* [Online]. Available at: http://arxiv.org/abs/1206.2682.ArXiv:1206.2682 [Accessed January 2021].

Frigg, R. (2004). In what sense is the Kolmogorov-Sinai entropy a measure for chaotic behavior? Bridging the gap between dynamical systems theory and communication theory. *British Journal for the Philosophy of Science*, 55, 411–434.

Gàcs, P., Hoyrup, M., Rojas, C. (2011). Randomness on computable probability spaces – A dynamical point of view. *Theory Comput. Syst.*, 48(3), 465–485 [Online]. Available at: http://dx.doi.org/10.1007/s00224-010-9263-x [Accessed January 2021].

Galatolo, S., Hoyrup, M., Rojas, C. (2010). Effective symbolic dynamics, random points, statistical behavior, complexity and entropy. *Information and Computation*, 208(1), 23–41 [Online]. Available at: http://www.sciencedirect.com/science/article/pii/S0890540109 001461 [Accessed January 2021].

del Giudice, E. (2007). Old and new views on the structure of matter and the special case of living matter. *Journal of Physics: Conference Series*, 67, 012006.n [Online]. Available at: http://www.i-sis.org.uk/Emilio_Del_Giudice.php [Accessed January 2021].

Gould, S. (1989). *Wonderful Life*. Norton, New York, USA.

Gould, S. (1997). *Full House: The Spread of Excellence from Plato to Darwin*. Three Rivers Press, New York, USA.

Graham, R. and Spencer, J.H. (1990). Ramsey theory. *Scientific American*, 262, 112–117 [Online]. Available at: http://dx.doi.org/10.2307/2275058 [Accessed January 2021].

Hilbert, D. (1930). Naturerkennen und logik naturerkennen und logik, 230 [Online]. Available at: http://www.jdm.uni-freiburg.de/JdM_files/Hilbert_Redetext.pdf [Accessed January 2021].

Kochen, S.B. and Specker, E. (1967). The problem of hidden variables in quantum mechanics. *Journal of Mathematics and Mechanics* (now *Indiana University Mathematics Journal*), 17(1), 59–87 [Online]. Available at: http://dx.doi.org/10.1512/iumj.1968.17.17 004 [Accessed January 2021].

Kupiec, J. (1983). A probabilistic theory of cell differentiation, embryonic mortality and DNA c-value paradox. *Specul. Sci. Techno.*, 6, 471–478.

Kupiec, J.J. (2010). On the lack of specificity of proteins and its consequences for a theory of biological organization. *Progress in Biophysics and Molecular Biology*, 102, 45–52.

Kwon, O.H. and Zewail, A.H. (2007). Double proton transfer dynamics of model DNA base pairs in the condensed phase. *Proceedings of the National Academy of Sciences*, 104(21), 8703–8708 [Online]. Available at: http://www.pnas.org/content/104/21/8703 [Accessed January 2021].

Laloë, F. (2012). *Do We Really Understand Quantum Mechanics?* Cambridge University Press, Cambridge [Online]. Available at: www.cambridge.org/9781107025011 [Accessed January 2021].

Laplace, P.S. (1998). *Philosophical Essay on Probabilities*. Translated from the 5th French edition of 1825. Springer, Berlin, New York [Online]. Available at: http://www.archive.org/details/philosophicaless00lapliala [Accessed January 2021].

Laskar J. (1994). Large scale chaos in the solar system. *Astron. Astrophys.*, 287, L-L12.

Longo, G. (2012). Incomputability in physics and biology. *Mathematical. Structures in Comp. Sci.*, 22(5), 880–900 [Online]. Available at: http://dx.doi.org/10.1017/S096012951 1000569 [Accessed January 2021].

Longo, G. (2017). How future depends on past histories in systems of life. *Foundations of Science*, 1–32 [Online]. Available at: http://www.di.ens.fr/users/longo/files/biolog-observ-history-future.pdf [Accessed January 2021].

Longo, G. (2018). Interfaces of incompleteness. In *Systemics of Incompleteness and Quasi-systems*, Minati, G., Abram, M., Pessa, E. (eds). Springer, New York.

Longo, G. and Montévil, M. (2014a). *Perspectives on Organisms: Biological Time, Symmetries and Singularities*. Springer, Berlin and Heidelberg.

Longo, G. and Montévil, M. (2014b) *Perspectives on Organisms: Biological Time, Symmetries and Singularities*. Lecture Notes in Morpho-genesis. Springer, Dordrecht.

Longo, G. and Montévil, M. (2015). Models and simulations: A comparison by their theoretical symmetries. In *Springer Handbook of Model-Based Science*, Dorato, M., Magnani, L., Bertolotti, T. (eds). Springer, Heidelberg [to appear].

Longo, G., Montévil, M., Kaufman, S. (2012a). No entailing laws, but enablement in the evolution of the biosphere. In *Genetic and Evolutionary Computation Conference, GECCO'12*. ACM, New York [Invited Paper].

Longo, G., Miquel, P.A., Sonnenschein, C., Soto, A.M. (2012b). Is information a proper observable for biological organization? *Progress in Biophysics and Molecular Biology*, 109(3), 108–114.

Longo, G., Montévil, M., Sonnenschein, C., Soto, A.M. (2015). *In Search of Principles for a Theory of Organisms*. Journal of Biosciences, Springer, 40(5), 955–968.

Luo, Z.X. (2011). Developmental patterns in mesozoic evolution of mammal ears. *Annual Review of Ecology, Evolution, and Systematics*, 42, 355–380.

Marinucci, A. (2011). *Tra ordine e caos. Metodi e linguaggi tra fisica, matematica e filosofia*. Aracne, Rome.

Monod, J. (1970). *Le hasard et la nécessité*. Le Seuil, Paris.

Munsky, B., Trinh, B., Khammash, M. (2009). Listening to the noise: Random fluctuations reveal gene network parameters. *Molecular Systems Biology*, 5, 318–325.

Myrvold, W.C. (2011). Statistical mechanics and thermodynamics: A Maxwellian view. *Studies in History and Philosophy of Science Part B: Studies in History and Philosophy of Modern Physics*, 42(4), 237–243 [Online]. Available at: http://dx.doi.org/10.1016/j.shpsb.2011.07.00 [Accessed January 2021].

Novick, A. and Weiner, M. (1957). Enzyme induction as an all-or-none phenomenon. *Proceedings of the National Academy of Sciences*, 43(7), 553–566 [Online]. Available at: http://www.pnas.org/content/43/7/553.shor.

O'Reilly, E.J. and Olaya-Castro, A. (2014). Non-classicality of the molecular vibrations assisting exciton energy transfer at room temperature. *Nat. Common.*, 5 [Online]. Available at: http://dx.doi.org/10.1038/ncomms4012 [Accessed January 2021].

Pironio, S., Acin, A., Massar, S., de la Giroday, A.B., Matsukevich, D.N., Maunz, P., Olmschenk, S., Hayes, D., Luo, L., Manning, T.A., Monroe, C. (2010). Random numbers certified by Bell's theorem. *Nature*, 464(7291), 1021–1024 [Online]. Available at: http://dx.doi.org/10.1038/nature09008 [Accessed January 2021].

Poincaré, H. (1902). *La Science et l'hypothèse*.

Richards, E.J. (2006). Inherited epigenetic variation revisiting soft inheritance. *Nature Reviews Genetics*, 7(5), 395–401.

Shanahan, T. (2012). *Evolutionary Progress: Conceptual Issues*. John Wiley & Sons Ltd, Chichester.

Shapiro, J.A. (2011). *Evolution: A View from the 21st Century*. FT Press, Upper Saddle River, New Jersey.

Soifer, A. (2011). Ramsey theory before Ramsey, prehistory and early history: An essay in 13 parts. In *Ramsey Theory, Progress in Mathematics*, Vol. 285, Soifer, A. (ed.). Birkhäuser, Boston [Online]. Available at: http://dx.doi.org/10.1007/978-0-8176-8092-3_1 [Accessed January 2021].

Turing, A.M. (1950). Computing machinery and intelligence. *Mind*, 59(236), 433–460.

Turing, A.M. (1952). The chemical basis of morphogenesis. *Philosophical Transactions of the Royal Society of London. Series B, Biological Sciences*, 237(641), 37–72 [Online]. Available at: http://rstb.royalsocietypublishing.org/content/237/641/37 [Accessed January 2021].

Weihs, G., Jennewein, T., Simon, C., Weinfurter, H., Zeilinger, A. (1998). Violation of Bell's inequality under strict Einstein locality conditions. *Physical Review Letters*, 81, 5039–5043 [Online]. Available at: http://dx.doi.org/10.1103/PhysRevLett.81.5039 [Accessed January 2021].

Zeilinger, A. (2005). The message of the quantum. *Nature*, 438, 743 [Online]. Available at: http://dx.doi.org/10.1038/438743a [Accessed January 2021].

2

In The Name of Chance

Hasard: French [masculine noun] (chance). This word of Arabic origin (*az-zahr*) for dice first appeared in French via the Spanish "*azar*"[1]. Initially it referred to the game of dice before becoming the general word for an unforeseeable event, without an apparent cause (*"les hasards de la vie"*, life's chances) and, by extension, the mode of occurrence for events of this type (*"en passant par hasard"*, happening by chance).

2.1. The birth of probabilities and games of chance

In fact, the Arabic word for dice (*az-zahr*) is not only the etymological origin of the French word for chance (*hasard*), but also directly the origin of probability calculus. Indeed, courtiers of the 16th and 17th centuries were so royally bored that, in order to kill time in Tuscany or Versailles, people played dice. For money, of course, otherwise it would have just added to the boredom. As a result, some started to think about ways to outsmart their opponents and so devised tricks in the form of strategies, which, after all is said and done, are simply the first encounters with probability problems.

– The problem sent by the Prince of Tuscany to Galileo (1554–1642): Why, although there were an equal number of six ways to write the numbers nine and ten as the sum of three non-zero integers, did experience state that the chance of throwing a total nine with three fair dice was less than that of throwing a total of ten? (see (Hombas 2004) for an extension of this celebrated problem).

– The problem sent by the knight Chevalier de Méré (1610–1685) to his friend Blaise Pascal (1623–1662): is it more probable that one will roll at least one six after four throws of a single die, or to roll at least one double six after 24 rolls of two dice?

Chapter written by Gilles PAGÈS.

1 As is often the case, the precise etymology remains debated, but, as the CNTRL (*Centre National de Ressources Textuelles et Lexicales*) website indicates, the various alternative origins considered all revolve around the game of dice.

Besides the historical links between probability and games of chance, this highlights the fact that probability theory, as a mathematical discipline, arose relatively late, in comparison to other branches such as geometry, arithmetic and, to a certain extent, measure theory and analysis (the theory of functions).

2.1.1. *Solutions*

Readers with a greater passion for chance over combinatorial calculations can content themselves with reading the results below to reflect on their significance before moving on to section 2.1.2.

2.1.1.1. *The Prince of Tuscany's problem for Galileo*

When one rolls three dice (that are implied to be unloaded and indistinguishable), as the Prince of Tuscany noted, there are many ways to write 9 and 10 as the sum of the three dice. Indeed, this can be symbolized by an increasing triplet of three numbers lower or equal to 6 whose sum is, respectively, 9 or 10:

$$
\begin{aligned}
9 &= 1 + 2 + 6 \\
&= 1 + 3 + 5 \\
&= 1 + 4 + 4 \\
&= 2 + 2 + 5 \\
&= 2 + 3 + 4 \\
&= 3 + 3 + 3
\end{aligned}
\qquad \text{and} \qquad
\begin{aligned}
10 &= 1 + 3 + 6 \\
&= 1 + 4 + 5 \\
&= 2 + 2 + 6 \\
&= 2 + 3 + 5 \\
&= 2 + 4 + 4 \\
&= 3 + 3 + 4
\end{aligned}
$$

The possible results when rolling three dice can be represented as (ordered) triplets of three numbers between 1 and 6. There are thus, according to this way of counting, $6^3 = 216$ possible outcomes. Note that by doing this, each die corresponds to a "position" within the triplet.

– In the case of a sum equal to 9, we therefore obtain:

$$
\mathbb{Proba}\big(\text{sum is } 9\big) = \frac{6 + 6 + 3 + 3 + 6 + 1}{216} = \frac{25}{216} \approx 0.116.
$$

Indeed, if each of the three numbers of the sum are distinct, there are three possible choices for the first die, two for the second and one for the third, i.e. $6 = 3 \times 2 \times 1$ in order to obtain 9 as the sum of these three numbers. If the sum is obtained as the sum of two distinct numbers, one being used twice, there is only a choice of the position of the unrepeated number, that is, three possible choices. If 9 is obtained as the sum of three times the same number, there is only one way to do it. Hence, the result is obtained.

– In the case of 10, we thus obtain, through the same arguments:

$$\mathbb{P}\text{roba}(\text{sum is } 10) = \frac{6+6+3+6+3+3}{216} = \frac{27}{216} \approx 0.125.$$

The Prince of Tuscany was therefore right! We deduce, in passing, that he had an instinctive intuition for the law of large numbers (LLN; see below) and, above all, a lot of free time to validate this through experiments...

2.1.1.2. *Chevalier de Méré's problem for Pascal*

When rolling four (unloaded) dice, there are 6^4 possible outcomes, all equally probable. It is easier to count the throws without any sixes, that is to say, the probability of this happening, almost immediately:

$$\mathbb{P}\text{roba}(\text{at least one six}) = 1 - \mathbb{P}\text{roba}(\text{no sixes}) = 1 - \frac{5 \times 5 \times 5 \times 5}{6^4}$$

$$= 1 - \frac{625}{1296} \approx 0.5177.$$

We roll two dice: $6^2 = 36$ possible outcomes. We roll these two dice 24 times so that there are $36^{24} \approx 2.245226 \times 10^{27}$ possible outcomes (equally probable, if the dice are not loaded).

Thus[2]:

$$\mathbb{P}\text{roba}(\text{at least one double-six}) = 1 - \mathbb{P}\text{roba}(\text{no double-six})$$

$$= 1 - \frac{35^{24}}{36^{24}} \approx 0.4914.$$

The LLN is once again hiding within this question, in the same deceptively disinterested manner as with the previous problem.

In both cases, these problems of two probability events turned out to be very close to the *random* dice roll tests, the implication being that if one was to repeat this a great many times, one would observe – discreetly – that the most probable event would occur with greater frequency. No doubt, equipped with this knowledge, it certainly would have been possible to seize the advantage.

---

2 Let us point out for the record that, given the orders of magnitude involved, Pascal very likely made use of the first logarithm tables in order to finalize his calculation, which, in a bout of luck, the Scottish mathematician John Neper (or Napier, 1550–1617) had discovered and tabulated a few years earlier, in 1614. Or else, which is more likely, he used the *Pascaline*, the calculating machine he had invented at the age of 19 (in 1642).

2.1.2. *To what end?*

Armed with an exact and precise numerical answer to their questions, the prince and the knight could place bets in their respective courts against other players. By systematically betting on the most probable configuration, the LLN (and time...) allowed them to benefit financially from the slight bias to which only they were aware (aside from their mathematical advisers, who were probably not around). The ideas for such gambling strategies highlights an intuition for the notion of *mathematical expectation*, a quantity which, in a game of chance, is expressed as the product of the winnings and their probability of occurrence, summed over all possible winnings.

In fact, the games of chance were likely behind the more profound precursors of probability mathematics that took place between Pascal and Fermat (1601–1665), who, from 1654 onwards, began a long correspondence, initially motivated by the problem of the distribution of stakes, which is broadly considered to be the first theoretical contributions to probability theory. This problem was much debated at the time, again for the same reasons, and consisted of determining a fair distribution of bets among those participants taking part in the event, when the game of chance in question was halted prematurely. It was Pascal who first submitted to Fermat a solution that he had devised. Several other eminent scholars at the time offered their contributions, all of which had turned out to be incorrect. In modern language – if one thinks in term of averages – the answer is immediate: the bets must be distributed in proportion to the probabilities that each of the participants had of winning at the moment of interruption. The problem was that, at the time, the concept of probability, and, especially, that of expectation, remained to be invented or, at least, precisely defined. Indeed, in this correspondence, published in 1679, the notions of probability (as a number between 0 and 1), but also that of mathematical expectation, are clearly being used, but without ever being extracted from the original context.

Today, we would say that Fermat's reasoning was more akin to that of a *combinatorist*, counting favorable cases and possible cases, whereas Pascal, a "disciple of betting" and the inventor of mathematical expectation, is more inclined to Game Theory. The precise origin of *mathematical expectation* is to be found in the famous text of Pascal's *wager* (see Pascal 1897, Pensée 223), the aim of which was to demonstrate to the libertines that they had a *vested interest* in believing in God. It was on this occasion that the philosopher and mathematician introduced the founding concept of mathematical expectation, consisting of the multiplication of a gain (in this case, the finite quantity of earthly pleasures versus the infinite quantity of eternal pleasures of a paradisiacal life) by the probability that God exists, which one can assume to be arbitrarily small, but which likewise one cannot positively assert to be nil, regardless of the intensity of one's own disbelief.

Discovering that the origin of probabilities is in games of chance (similar to the way in which statistics paved the way for actuarial calculations) is a heavy burden for a science to bear, but for the greatest French probabilist of the 20th century it is a given. In 1970, a year before his death, Paul Lévy (1886–1971), a probabilist with phenomenal intuition and, in many ways, one of the most singular mathematicians of the 20th century, wrote:

> [...] what solid bodies are to geometry, games of chance are to probability calculus, but with one difference: solid bodies are given by nature, whereas games of chance were created to test a theory imagined by the human mind, so that the role of pure reason is even greater in probability calculus than in geometry. (Lévy 1970)

2.2. A very brief history of probabilities

The intention here is not to give a thorough history of probability, but to nonetheless provide an outline of the key stages, so that one might establish a timeline of the principle ideas. The story thus begins with the Pascal–Fermat correspondence that began in 1654. Next, as the first cornerstone, the (weak) LLN was established by Jacob Bernoulli (also known as *Jacques 1er*, 1654–1705) and published posthumously by his nephew Nicolas in the book *Ars conjectandi* in 1713 (Bernoulli 1713). This first version of the LLN was established on the framework of coin-tossing, the game of heads or tails (with a coin that is not loaded). Through combinatorial methods, Bernoulli demonstrates that when we toss a balanced coin *ad infinitum*, we observe, little by little, that the frequencies for heads and tails converge to $1/2$, the probability *a priori* of obtaining heads or tails. Various generalized conjectures of this nature have been put forward up until the modern version, which extends this behavior to the indefinite and independent repetition of any random phenomenon. However, if we think back to the Prince of Tuscany and the Chevalier de Méré, we understand that many others before Bernoulli not only had this intuition, but had already imagined how to take advantage of it.

The second stage of the probabilistic rocketship is what specialists call, in a particularly blunt way, the Central Limit Theorem. This is none other than the description of the speed at which empirical frequencies approach the *a priori* probability of the random event reproducing itself independently. It is a measure of fluctuations around the mean that turns out to follow a probabilistic law which is called normal distribution, but also Gauss's law or, sometimes, the Bell Curve. The word normal here testifies to the universal character of these fluctuations. It was Abraham de Moivre (1667–1754) who highlighted the phenomenon, again in the case of a (fair) coin toss game. Carl Friedrich Gauss (1777–1855) then generalized the result in the more general framework through the analysis of uncertainties in physical measurements.

Other major contributors to probability have appeared over the years, notably Laplace and Poisson, to name the Frenchmen, and Gauss, of course, but nevertheless we can mark the birth of modern probability theory at the beginning of the 20th century, with the work of Émile Borel, who applied Lebesgue's theories of measure and integration – authentic mathematical revolutions at the time – to probabilistic questions as early as 1900 (like normal numbers) and then, above all, the works of Andreï N. Kolmogorov, who, relying on these new theories, in *Grundbegriffe der Wahrscheinlichkeitsrechnung*, published in 1933 (Kolmogorov 1933), posed the axioms of probabilities (and stochastic processes) in a way that remains definitive to this day.

2.3. Chance? What chance?

Throwing dice can be argued as an experiment in solid mechanics. There is no chance involved, only a differential equation. With the exception of said equations' sensitivity to the initial conditions, it is impossible to predict the behavior of the dice, and therefore random modeling is that much more efficient.

In ancient times, the Babylonians, for example, considered eclipses and comets to be random phenomena, and yet today, any person today possessing a postal calendar can easily refute such claims. Privileged to be (almost) a sexagenarian, my birth was a random phenomenon for my parents: they did not know throughout the whole pregnancy what the sex of their child would be. However, thanks to the invention of echography, my own children were determined as girls and boys long before they were born. That being said, what could be known of their future sexual orientations?

On another issue, quantum mechanics has given rise to a titanic controversy as to its irreducibly random quality: does an electron have a "position" in the common sense of the term or is this, by nature, a probability distribution in space? Niels Bohr's so-called *Copenhagen Interpretation* points in this direction, to which Albert Einstein famously proclaimed: "[God] does not play dice"; implying the existence of a system of hidden variables with the sole purpose of hiding this element of *chance* that Einstein did not wish to see, to which, not to be outdone, Bohr replied: "Einstein, stop telling God what to." (In fact the controversy was more about the local nature of quantum mechanics). The experiment carried out by Alain Aspect and his team from the Institute of Optics in the 1980s, around the Einstein-Podolsky-Rosen paradox, would decide in favor of the former, even if some were never fully convinced, such as the French physicist Louis de Broglie. After this, some physicists began referring to *objective chance* as a way of considering that there is an irreducible part of chance that cannot be reduced to a simplified interpretation or model.

The contestation over the existence of objective chance has been the preserve of eminent physicists. Maurice Allais (1911–2010), the economist (and also something of a physicist, it is true) at the dawn of the 1990s, took advantage of the prestige of his Nobel Prize to fiercely contest the existence of objective chance, at the same time denouncing random modeling as dubious efficiency. Among these adepts of "dubious" chance, the mathematician René Thom, the father of Catastrophe Theory and a keen study of the human sciences, is yet another fierce adversary to randomness.

The field wherein each of us is able to ask ourselves this question is the life sciences. In particular, the case of sexual reproduction and its processions of spermatozoa and egg cells. There are some who like to think of it as a race that each of us has won, and as such everyone has, at least once, been a winner, thus the lyrics of the popular French song:

> *De Ruth ou de Moïshé, lequel a eu l'idée ? Qu'importe j'ai gagné la course, et parmi des milliers Nous avons tous été vainqueurs, même le dernier des derniers, Une fois au moins les meilleurs, nous sommes nés.*

> (From Ruth or Moses, whose idea was it?

> It doesn't matter, I won the race, and out of thousands We were all winners, even the last of the last, The best, for one moment, at least, before we are born.)

> (Bonne idée, J.J. Goldman/J.J. Goldman, *"En passant"*, Columbia, 1998)

However, for geneticists, this event is the complete opposite of competition: it is a lottery patiently designed and organized by nature through sexual reproduction, with the function of maintaining the species' genetic pool. The genetic makeup of a human being is composed of 23 pairs of chromosomes. Each sex cell (sperm, egg) contains 23 chromosomes (one from each pair). A couple can therefore produce:

$$2^{23} \times 2^{23} = 2^{46} \approx 7.0368 \times 10^{13} \simeq 70,368 \text{ billion combinations.}$$

On this crazy merry-go-round, it is no longer a question of solid mechanics and chaotic differential equations, but rather the outcome of this or that chromosome, deciding the sex or color of the eyes of the yet-to-be-conceived child. There are so many factors that are likely to impact the outcome that any deterministic modeling of the phenomenon becomes illusory, regardless of any hypothetical knowledge or calculation skills yet to be introduced. Hence, the familiar trope of *"life's lottery"*, which, in fact, brings us back to the games of chance, the aforementioned historical basis of probability theory, according to Paul Lévy. To cut short these digressions, we will qualify this lottery as an *effective* method for randomness according to the terminology proposed by the physicist Hubert Krivine.

In the reflections and controversies briefly mentioned above, we find few, if any, probabilists (mathematicians), especially once Kolmogorov published his seminal work, in which the word "chance" is largely absent and does not give rise to any particular development as to its existence and underlying nature.

In fact, Kolmogorov's very abstract axioms appear to be a metamodel within analysis (especially for measure theory and integration, as mentioned above) whose specificity is reflected, in particular, through a precise definition of the independence of events. Upon this incredibly powerful conceptual foundation, the whole theory of probability and stochastic processes, both frequentist and Bayesian mathematical statistics, has been developed. Kolmogorov's axioms allow one to model and analyze, with the same efficiency, very disparate random phenomena, whether they are the result of objective or *effective randomness*, or even for an incomplete simple information system (partial observation). The relationships that actors in the sciences studying randomness (probability theory, statistics) have with chance, its nature and its sources, have much to do with the genesis of their disciplines as metamodels within mathematics. Caricaturing a little, probabilists and statisticians have a definite tendency to validate their hypotheses on the stochasticity of a phenomenon through the theorems they derive from it, rather than to focus on the nature of chance as described by said hypotheses. The predictive capacity of the theory, in turn, validates the hypotheses made. This brings us back to the beginning as this approach is itself almost a definition of modeling in science.

This point of view seems perfectly illustrated by Paul Lévy who, in 1940, wrote in (Lévy 1940):

> [...] in a concrete case, in Brownian motion for example, nature only realizes phenomena on a microscopic scale that the mathematician, for the convenience of his research, extends to the infinitely small. Moreover, we believe – although eminent scientists have not been of this opinion since the work of Heisenberg – that the notion of chance is a notion the scientist introduces because it is convenient and fertile, but one which nature ignores.

Paul Lévy is in fact taking part in the Copenhagen controversy, pitting himself against the existence of objective chance, even though he has devoted his life and energy to the study of random phenomena! Proof that, in his esteemed opinion, chance and its ontological nature are not a worthy subject *per se*, and, in any case, not sufficiently relevant enough to distract him from his probabilistic mathematical activities.

Let us add that if quantum mechanics has become the archetype of objective chance, it is not however a matter of "probabilistic" chance, if by which we mean Kolmogorov's axioms. Therefore, the notion of "classical" independence does not really play a central role within it. The chronology undoubtedly plays an important part in this plan, but is not the only factor.

Without forgetting the humorous approach:

> Chance was created by God, who, not wishing to be in charge of everything, invented it to convince men of the contrary. Unless it is the other way around.

It should be noted that, as mentioned above, these discussions and these controversies between eminent scientists almost never invite probabilists to the table; these same probabilists themselves are likewise careful not to set foot in side the door.

2.4. Prospective possibility

There is, however, a third category of chance, one which the human being (in the role of *demiurge*) has created and placed in the depths of computers, which is (almost) the very purpose for which computers were created.

2.4.1. *LLN + CLT + ENIAC = MC*

The idea of instrumentalizing chance in pursuit of other objectives germinated in the human mind long before the ENIAC (Electronic Numerical Integrator And Computer), which was the first real "Turing-complete" computer, built in 1945. As early as 1733, Georges-Louis Leclerc, Comte de Buffon and pre-Darwinian naturalist, proposed to the *Académie des Sciences* (see Buffon 1733) a method for the approximation of π that consisted of throwing a needle of length a onto a parquet floor with parallel slats of width d, with $a < d$, a great number of times. The frequency with which the needle overlaps two slats tends toward the theoretical probability $\frac{2a}{d\pi}$, allowing us, if one knows a and d, to estimate π. Beyond the subtle calculation, which over the years has become an essential exercise in probability, Buffon was the first to calculate a quantity through probabilistic simulation. This went well beyond a clear intuition, or even his likely knowledge of the (so-called "strong") LLN. He understood how to frame the variables for it to be an effective method of calculation. In particular, he assumed that the throws, notably the successive angles of incidence of the needle on the parquet slats, were independent. The method is, unfortunately, notoriously inefficient because it is far too slow at calculating decimal places of π.

In fact, the real story of this instrumentalization of chance begins in 1946, out of the Manhattan Project at the Los Alamos laboratory under the direction of John von Neumann, who was tasked with developing the first atomic bomb. Stanislaw Ulam (1909–1984), a Polish-born mathematician and physicist who fled to the United States, had been called in to work on the project by John von Neumann, himself an American of Hungarian origin.

One evening while playing a game of "patience" (*Canfield solitaire*), Ulam took to calculating the theoretical probability of winning but quickly encountered the combinatorics of the problem. This sparked the idea of *simulating a great many games* on the newly born ENIAC computer and thus calculate the desired probability via the ratio between the number of games won over the number of games simulated.

As a result of this, the *Monte Carlo* (MC) *method* was born, that is to say, the experimental implementation of the strong LLN through computer simulation, in combination with the Central Limit Theorem (CLT), which controls the speed of convergence that allows for the design of a confidence interval by a variance calculation[3].

von Neumann immediately recognized the scope of the idea, which, beyond the probabilities of an event, makes it possible to calculate the average – (mathematical) expectation – of any random phenomenon that can be simulated numerically. Given the context of his situation, he naturally thought of its application to the numerical solving of partial differential equations in Neutronics (a special case of the Boltzmann equation) "through" the representation of their solutions as mathematical expectations of particle phenomena. In 1949, Nicholas Metropolis and Ulam went on to publish their seminal paper, The Monte Carlo Method ((Metroplis and Ulam 1949), the general focus of which can be said to be concerned more with physics than mathematics). Incidentally, why did they choose to name it after "Monte Carlo"? This was the – highly confidential – code name for the project and N. Metropolis would later account for it being thus named after the proclivities of Ulam's uncle, a loyal player at the eponymous casino. Fermi, who independently came to similar conclusions, albeit through different routes, would later join the team and begin an ongoing collaboration (the Fermi-Pasta-Ulam experiment) with Ulam. The major short-term consequence was the development, by Teller and Ulam, of the H-bomb (for which a patent was filed!) in 1952. The MC simulation played an important role, in particular, helping to resolve a few discrepancies between the two researchers along the way.

2.4.2. *Generating chance through numbers*

One major obstacle remained: how to generate random numbers, which is the essential input for any effective implementation. This vast subject quickly came to occupy the team. Several proposals were put forward, including those now famous for being incorrect. Two paths were followed: pseudo-random numbers imitating, as closely as possible, numbers drawn at random, uniformly and independently,

---

3 This is an interval centred on the calculated value within which lies the true value being sought, with a probability fixed in advance that is generally between 95% and 99.99%. Its size decreases inversely to the square root of the number of simulations (or trials).

between 0 and 1; and quasi-random numbers that, sacrificing certain truly "random" properties, made it possible to accelerate the convergence of the LLN. The second option, defended, in particular, by T. Warnock and which attracted von Neumann for a time, led to the method known as the "Quasi-Monte Carlo", which went on to be developed further by many authors (Kuipers, Halton, Niedderreiter and a whole Austrian school, which has taken over the project up to the current day), imbuing solid theoretical foundations and connecting it with ergodic theory and dynamic systems. However, in practice, it ultimately suffers from a dependence on an intrinsic dimension of the phenomenon that limits its performance. The *intrinsic dimension* referred to is the (average) quantity of random numbers necessary to simulate one scenario of the phenomenon.

The first approach, whose foundations are less certain, has given rise to numerous works and controversies, emanating from communities of logicians and computer scientists, rather than probabilists, who, once again, were quite quick to frown upon the subject as poorly postured: how to imitate chance?

The solutions put forward by these communities were somewhat disconcerting: the logician severs this Gordian Knot by stipulating that any series of numbers, the "fruit" of an algorithm, is not "by chance"; this is the case with the decimals of π, which some see as the archetype for a sequence of random numbers. To the computer scientist, who is hardly more accommodating to applied mathematics than to simulation, chance is but a series of numbers whose simulation of the first n terms requires a code whose length is proportional to n.

Finally, is it not better to fall back on a definition Émile Borel put forward in his work "*Le Hasard*" ("*Chance*" (Borel 1920)), which introduced normal numbers (of $[0, 1]$) characterized by an uniformly distributed n-adic expansion over $\{0, \dots, n-1\}$ in any base $n \geq 2$? Almost any real number in $[0, 1]$ is "normal", however we do not know of any that does not fall under the condemnation of computer scientists (and therefore logicians).

It was not with these kind of theoretical errors that we were going to solve the simulation of neutronic equations and, as a result, the more efficient manufacture of hydrogen bombs. It was necessary to settle the matter, since the MC method only makes sense if we can generate very large quantities of random numbers very quickly, especially given that, at the time, the storage of information was even more challenging than the calculation of it. After some trial and error, wherein eminent mathematicians talked a great deal of nonsense, congruent generators, based on elementary arithmetic in $\mathbb{Z}_n$ sets and where calculations are made to be modulo n (for example: $1 + 1 = 0$ modulo 2; $7 \times 3 = 1$ modulo 10; etc.), came to the rescue of practitioners, where n is set as a very large prime number. The search for these very large prime numbers found an unexpected application, and was soon incorporated into cryptography, but that is another story. It was nevertheless necessary to wait for these pseudo-random number generators to be made more reliable, in particular by significantly increasing the time period.

Later on, through its application to finite fields (register transfers), the situation effectively stabilized and today reliable pseudo-random number generators are accessible directly. The *Mersenne Twister Generators* (Matsumata and Nishimura 1998) are the most commonly used by the computational libraries of typical computer languages[4]. To further investigate these congruential generators, we refer to section 2.5., the Appendix, which explains in greater detail how it was devised.

2.4.3. *Going back the other way*

In the decades that followed, physicists were the principle users and developers of the MC simulation, which was regularly called upon to calculate various physical constants and various regimes of phase transition for a variety of fields.

However, another playing field loomed on the horizon, the origins of which can be found at the very beginning of the 20th century, with the first modeling of stock market prices by Louis Bachelier, in his thesis, which he defended at the *École Normale Supérieure* under the (moderately benevolent) direction of Henri Poincaré (Bachelier 1900). These pioneering works developed a link between the heat equation – the simplest Parabolic Partial Differential Equation (PDE) – and the evolution of stock market prices (supposedly governed, we would currently say, by Brownian motion and which made their first "mathematical" appearance here), and in doing so, Bachelier deduced a means by which to calculate the price of certain financial products. Nevertheless, this work, too far ahead of its time, had no direct application to the markets and thus fell into partial obscurity. Seventy years on and two world wars later, the deregulation of financial markets radically changed this situation by authorizing the creation of organized exchanges, which permitted the trading of derivatives (futures, options, etc.), bringing Bachelier's work back into the spotlight, albeit posthumously.

Indeed, from the 1973 publication of the groundbreaking article by Fisher Black and Myron Scholes (Black and Scholes 1973) on the valuation and hedging of stock options traded on organized exchanges, such as the Chicago Board of Trade (CBOT), which had opened just a few months earlier, a shiver was sent throughout the financial

---

4 Note, however, that in the 1970s, NASA tried to impose a standard in this field by proposing a generator-type (debugging a program comprising of a random simulation or reproducing a result obtained by this means is a nightmare if one has no "references"), but it turned out that this simulator had a weakness, intentionally introduced to allow for a form of reverse engineering. A spy generator, that is, the precursor to computer viruses. (For further reading, see the *New York Times*).

world: it suddenly became imperative to learn and understand Brownian motion and the associated *stochastic differential calculus*, since, in this formula, the return on a stock is modeled by Brownian motion. Now, the price – or premium – of an option is an expectation, or, if one prefers, an infinitely weighted average over all possible scenarios of price variations. However, in the very simple case of a basic call option on a stock that depends only on the price of that stock at a given date in the future[5], the two researchers upset the calculation of probabilities, deriving the famous Black–Scholes–Merton formula, not by calculating this expectation, but via the (explicit) resolution of a PDE that occurs naturally through their approach, driven by the search for infinitesimal hedging. It was Michael Harrison and Stanley Pliska, influenced by the ideas of David Kreps, who, a few years later, in 1981 (Harrison and Kreps 1979, 1981), made option premiums explicit in the form of an expectation.

In order to do this, while developing, in passing, the notion of infinitesimal hedging (or *δ-hedge*), which was their decisive contribution, when compared to Bachelier's work, they again used a representation formula – the Feynman–Kac formula this time – to establish that this expectation is the solution of a parabolic PDE whose links to the heat equation allow for an explicit resolution using elementary functions that are directly available by computer[6].

Their approach is therefore exactly opposite to that of Ulam, Metropolis, von Neumann or Fermi, who, for their part, solved a neutronics equation through the Monte Carlo simulation, after having established that its solution represented itself as an expectation. However, like any paradox, this only becomes apparent after the fact.

The existence of an explicit solution (closed form) is a welcome "accident" in the Black and Scholes article: PDEs are usually solved through numerical approximation schemes. If these strategies are the true Ferraris of low dimensional computations (options on one, two, or even three "underlying" stocks in finance), they quickly suffer when the number of underlying assets increases beyond that. One must then return to the approach developed by Ulam and his colleagues, since only the Monte Carlo method can play at that level. The use of simulation to compute option premiums and hedge multiple underlying products was, in fact, very quickly made compulsory among practitioners, several years prior to the first academic publications (Boyle 1977).

---

5 $\mathbb{E}\left[\max(S_T - K, 0)\right]$ where $S_T = s_0\, e^{-\frac{\sigma^2 T}{2} + \sigma W_T}$ and W_T denotes a standard Brownian motion at the date T, K the exercise price, s_0 the price of the stock at the moment 0 and S_T its (random) price on the date T.

6 In fact, the notion of expectation does not even appear as such in the article by Black and Scholes: the hedge is obtained as the result of reasoning on the absence of an arbitrage opportunity on a stochastic differential equation and the premium by resolution of a PDE via the Feynman–Kac formula.

This new application quickly turned out to be a real makeover for the MC simulation, sparking a great deal of research in order to improve its efficiency and, especially, the speed of convergence. This was aided by the acceleration of computer calculation speeds, in accordance with Moore's law, until it gave rise to the new discipline of numerical probability, the object of which is to generally improve upon the efficiency of probabilistic numerical methods of calculation and simulation, such as, in addition to the MC simulation, its various variants: the Quasi-Monte Carlo, the Metropolis-Hasting, the Markov Chain Monte Carlo, etc., up to and including stochastic optimization, since the history of prospective possibility will not end with "simple" simulation.

2.4.4. *Prospective possibility as master of the world?*

If one were to roll a lead marble from the top of Everest along a ledge, there is a very small chance that it would later be found in the depths of the Mariana Trench, the lowest point on the planet, assuming that there was a downward path from one to the other. This is because the marble runs the risk of being trapped at a *local lowest point* along the way, that is to say, a point from which it would need to "climb out". As an introductory cliff-hanger this is admittedly a little bombastic, so let us rather think, more prosaically of a large warped box containing a minimum of 100 eggs, each one individual in its shape, and remake the same experiment. The conclusion will be the same and, better yet, observed inexpensively. This is the problem encountered by any optimizer (human or not) who wishes to find the minimum of a (regular) function f depending on a great number (how great? In the previous example it was only $d = 2 \dots$) of variables $x_1, \dots, x_d, d$. The ball drop discussed above is (distantly) akin to the so-called *gradient descent* method.

The idea behind most stochastic optimization methods is to, to a greater or lesser extent, naturally make this *gradient descent* go up the slopes from time to time, in a random way, so as to allow it to explore horizons other than the basin of attraction for a "local" minimum closest to the starting point and toward a standard gradient descent that will inevitably converge. By allowing such locally counterproductive behavior, we give ourselves a greater or lesser chance of hitting upon the Holy Grail of the absolute minimum. From gradient descent to simulated annealing to genetic algorithms (which dispense with derivative functions), all stochastic optimization algorithms are based on this paradigm. The oldest – developed in 1951 as part of a research program supported by the US Navy (Robbins and Monro 1951) – and the least sophisticated is the *Stochastic Gradient Descent* (SGD), which, after years of being eclipsed, seems to be the essential compromise between efficiency and implementability. In any case, it is the all-weather optimization tool that accompanied the emergence of *deep learning* for "neural networks" in artificial intelligence (AI) at the dawn of the 2000s. Born in the 1950s with the beginnings of computing, eclipsed in the 1970s, it experienced its first comeback in the 1980s, before today's explosion through, in particular, image recognition.

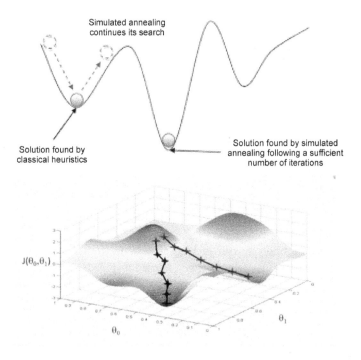

Figure 2.1. *Simulated annealing (above)*
and stochastic gradient (below)

   The optimization of such a supervised learning network consists of comparing, for each data input, the machine's response to the exact response, and in the case of error, correcting the network by modifying its parameters, which are, in fact, the strengths of the links between its microcomputing units. These links, which numbered in the tens of thousands in the networks of the 1980s, have reached several million, and, even more recently, over a billion (*Google* according to *ImageNet* in 2013). These staggering numbers correspond to the mathematical dimension of the optimization problem and when we see the complexity of a "landscape" in two-dimension we can only be afraid of what can happen in these stratospheric dimensions. Of course, these networks need to be "pruned" through various expertise and choices, *a priori*, but without changing the orders of magnitude. However, this optimization – referring here more to learning – works and gives results at efficiencies that were inconceivable 20 years ago. This efficiency has been popularized by the general public because of the performance of *AlphaGo*, a program designed by Google's subsidiary *Deepmind*, which has become the best *Go player* on the planet[7], but which is now the basis of many emerging technologies, from product recommendation to facial recognition.

Chance here is neither objective, because it is simulated and therefore imperfect, nor effective, in the sense of biologists. It is productive in the sense that it is an irreplaceable computation or optimization tool. It is even *prospective* as it allows us to tackle problems which, either by their complexity or their dimensional altitude, are inaccessible to any other method known to date. Skeptics may prefer the less flattering name of *auxiliary* chance to remind one that, in this business, it is only an instrument called upon to rescue a situation, that is, simply a tool in service of another objective.

Nevertheless, this chance is no less fascinating than the other two, because who would have thought that, in order to overcome the (hugely) great dimension, our best ally was chance? We must also admit that, ironically, there is an oxymoron in the expression "stochastic optimization".

However, in the view of chance as a model or as a tool, which suits probabilists and statisticians, the metaphysics of chance remains essentially absent from our conversations and mathematical exchanges. We are resolutely downstream. This does not prevent us from spending long days and a few nights trying to elucidate deeply random models (statistical mechanics, random maps, ballot box models, forest fires, risk calculations, etc.) or to analyze statistical estimators, both parametric and non-parametric. Randomness is formalized and modeled in the form of a probabilized space symbolized $(\Omega, \mathcal{A}, \mathbb{P})$ (or $(\Omega, \mathcal{A}, (\mathbb{P}_\theta)_{\theta \in \Theta})$ in Statistics), in which Ω represents the set of $\omega \in \Omega$ scenarios (or chances) possible, $\mathcal{A}$ represents a sort of coherent system of observable events and $\mathbb{P}$ measures the degree of likelihood of their occurrence. In the case of Statistics, there are as many possible statistical models as there are parameters θ. One would be astonished at the lack of attention paid to this triplet by probabilists and statisticians themselves, considering all that is built upon it.

But nothing is set in stone. As such, recent developments in the *Theory of Rough Paths and Regularity Structures* (Hairer 2014), for which the Fields medal was awarded to our colleague Martin Hairer in 2014, have a more ambiguous relationship with the notion of chance. This is enough to inspire our up and coming thoughts and reflections.

---

7 *AlphaGo* is essentially a very large artificial neural network, and Stochastic Gradient Descent is used in the learning phase as an optimization tool, including vast results of Monte Carlo simulations, in which the machine was playing against itself.

2.5. Appendix: Congruent generators, can prospective chance be periodic?

2.5.1. *A little modulo n arithmetic*

Let us pause for a moment and look at this notion of congruent generators, which illustrates both the incredible ingenuity and the fragility of this efficient process of random simulation. For once, we are going to have to get into the technicalities a little, not probabilistically as one might expect, but in terms of arithmetic. Indeed, everything depends on the properties of this seemingly elementary and yet deeply fascinating area of mathematics. However, it is not about the arithmetic of infinite sets of natural numbers $\mathbb{N} = \{0, 1, 2, \ldots\}$ or about relative integers $\mathbb{Z} = \{\ldots, -2, -1, 0, 1, 2, \ldots\}$, which everyone has knowledge of, but on finite sets $\mathbb{Z}_n = \{0, 1, 2, \ldots, n-1\}$[8], where n is a natural number greater than or equal to 2. To do arithmetic on this set, it is also necessary to have the two basic operations: addition and multiplication. To do this, we have to come back to $\mathbb{Z}$ and rely on *Euclidean division* or division with the remainder: the division which, when we divide 23 by 4, has 5 for the *quotient* and 3 for the *remainder*, since $23 = 5 \times 4 + 3$. Mathematicians formalize this operation as follows: let m be a natural number and n a non-zero integer: there then exists a unique *quotient* q in $\mathbb{N}$ and a unique *remainder* r in $\mathbb{Z}_n = \{0, 1, \ldots, n-1\}$, such that $m = nq + r$. Let us return to our operations on $\mathbb{Z}_n$. We define the addition $\oplus_n$ and a multiplication $\otimes_n$ between two elements a and b of $\mathbb{Z}_n$, therefore: we consider their sum $a + b$ and their usual product $a \times b$ in $\mathbb{N}$ which we both divide into integers through n. We therefore obtain:

$$a + b = n\,q_1 + r_1 \quad \text{and} \quad a \times b = n\,q_2 + r_2 \quad \text{where} \quad r_1, r_2 \in \mathbb{Z}_n$$

and we ask:

$$a \oplus_n b = r_1 \quad \text{and} \quad a \otimes_n b = r_2$$

We verify that these operations have all the desired properties, those observed in $\mathbb{N}$ by their usual counterparts ($a \oplus_n 0 = 0$, $a \oplus_n b = b \oplus_n a$, $a \otimes_n b = b \otimes_n a$, $a \otimes_n (b \oplus_n c) = a \otimes_n b \oplus_n a \otimes_n c$, among others). The addition $\oplus_n$ amends the good idea of behaving as if it is on $\mathbb{Z}$ since $a \oplus_n (n - a) = 0$ and $n = n \times 1 + 0$, which ensures that every element has an opposite for $\oplus_n$.

---

8 The most common arithmetic notation is $\mathbb{Z}/n\mathbb{Z}$, because it is actually a "quotient" set. I hope I do not shock the purists who come across this text... by chance.

We can draw up addition and multiplication tables for these operations. So, for $n = 3$:

$\oplus_3$	0	1	2
0	0	1	2
1	1	2	0
2	2	0	1

$\otimes_3$	0	1	2
0	0	0	0
1	0	1	2
2	0	2	1

and for $n = 4$:

$\oplus_4$	0	1	2	3
0	0	1	2	3
1	1	2	3	0
2	2	3	0	1
3	3	0	1	2

$\otimes_4$	0	1	2	3
0	0	0	0	0
1	0	1	2	3
2	0	2	0	2
3	0	3	2	1

Careful examination of these tables may leave the neophyte perplexed. Indeed, if $n = 3$, the subset $\mathbb{Z}_3^* = \{1, 2\}$ of $\mathbb{Z}_3^*$ has an unexpected property that $\mathbb{Z}$ itself does not have: 2 has an inverse for the multiplication in the sense that, like $2 \times 2 = 4 = 1 \times 3 + 1$, we have $2 \otimes_3 2 = 1$. It is even its own inverse. Therefore, the two elements of $\mathbb{Z}_3^*$ have an *inverse* of this multiplication (since $1 \otimes_3 1 = 1$, obviously) and the product of elements for $\mathbb{Z}_3^*$ remains in $\mathbb{Z}_3^*$. On the other hand, if $n = 4$, 3 does have an inverse itself (since $3 \times 3 = 2 \times 4 + 1$ and therefore $3 \otimes_4 3 = 1$), but 2 clearly does not have one (worse, $2 \otimes_4 2 = 0$!). We then verify that if we denote by $\mathbb{Z}_4^*$ the set of $\otimes_4$-invertible elements of $\mathbb{Z}_4$, then this does not match with non-zero elements of $\mathbb{Z}_4$, like when $n = 2$ or 3, but only consists of 1 and 3, and we therefore have $\mathbb{Z}_4^* = \{1, 3\}$.

If we further develop these investigations – with $n = 5$ – some elementary calculations show that, again, all elements of $\mathbb{Z}_5^* = \{1, 2, 3, 4, 5\}$ have an inverse of $\otimes_5$. Indeed:

$1 \otimes_5 1 = 1$

$2 \otimes_5 3 = 1$ because $2 \times 3 = 6 = 1 \times 5 + 1$

$3 \otimes_5 2 = 1$

$4 \otimes_5 4 = 11$ because $4 \times 4 = 16 = 3 \times 5 + 1$

whereas, in $\mathbb{Z}_6$, 3 does not have an inverse (check! But as $3 \otimes_6 2 = 0$, because $6 = 6 \times 1 + 0$ is fruitless...).

In fact, as soon as a natural integer n is a *prime*, that is to say, it has no other divisor than 1 and itself, as is the case with $2, 3, 5, 7, 11, 13, 17, 19, 23, 29, 31, \ldots$ then all the elements a of $\mathbb{Z}_n \setminus \{0\}$ have an inverse and we define $\mathbb{Z}_n^* = \{1, 2, \ldots, n-1\}$ the set of $(n-1) \otimes_n$-invertible elements of $\mathbb{Z}_n$. We therefore say that $\mathbb{Z}_n^*$ endowed with $\otimes_n$

is a commutative group (just like $\mathbb{Z}_n$ endowed with $\oplus_n$), which makes $\mathbb{Z}_n$ endowed with its two operations $\oplus_n$ and $\otimes_n$ a *field* (or *corps* in French, although sometimes a little abstruse, mathematicians are secretly poets).

Even better, we can show that – though much more delicate (see Perrin 1996) – when n is prime – there always exists at least one element a of $\mathbb{Z}_n$, such that:

$$\mathbb{Z}_n^* = \left\{ a^{\otimes_n k},\ k = 1, \ldots, n-1 \right\} = \left\{ a, a^{\otimes_n 2}, \ldots, a^{\otimes_n (n-1)} \right\}$$

where the $\otimes_n$-powers $a^{\otimes_n k}$ are defined step by step through

$$(*) \qquad a^{\otimes_n 1} = a, \quad a^{\otimes_n k} = a \otimes_n a^{\otimes_n (k-1)}$$

(and $a^{\otimes_n 0} = 1$ by convention). Such an integer a is called a *generator* of $\mathbb{Z}_n^*$ and $\mathbb{Z}_n^*$ endowed with $\otimes_n$ is called a *cyclic* or *monogenic* group.

By the way, where is 1 hiding in this enumeration? The answer is given by Fermat's (not so) small theorem, which ensures that $a^{\otimes_n (n-1)} = 1$, that allows for $\mathbb{Z}_n^*$, and it is also written as:

$$\mathbb{Z}_n^* = \left\{ a^{\otimes_n k},\ k = 0, \ldots, n-2 \right\} = \left\{ 1, a^{\otimes_n 1}, \ldots, a^{\otimes_n (n-2)} \right\}.$$

For example, in $\mathbb{Z}_5$, $a = 2$ is suitable because

$$2^{\otimes_5 1} = 2, \quad 2^{\otimes_5 2} = 4, \quad 2^{\otimes_5 3} = 3, \quad 2^{\otimes_5 4} = 1.$$

The $\otimes_5$-powers of 2 do exhaust $\mathbb{Z}_5^*$, but in a rather haphazard fashion. It's disorganized!

If we consider $n = 7$, we verify that:

$$5^{\otimes_7 1} = 5, \quad 5^{\otimes_7 2} = 4, \quad 5^{\otimes_7 3} = 6, \quad 5^{\otimes_7 4} = 2, \quad 5^{\otimes_7 5} = 3, \quad 5^{\otimes_7 6} = 1$$

or:

$$3^{\otimes_7 1} = 3, \quad 3^{\otimes_7 2} = 2, \quad 3^{\otimes_7 3} = 6, \quad 3^{\otimes_7 4} = 4, \quad 3^{\otimes_7 5} = 5, \quad 3^{\otimes_7 6} = 1.$$

Clearly, in these two enumerations of $\mathbb{Z}_7^*$, we go from disorganized to frankly erratic.

In fact, we can partially extend the preceding to the case where the integer n, without being prime, is a power $n = p^r$ of a prime number p.

We then speak of an integer that is *p-primary* or more simply *primary*. Indeed, more generally, when n is not prime, only the (non-zero) elements a of $\mathbb{Z}_n$ that are *prime with n*, i.e. having no common divisor with n other than 1, admit an inverse for $\otimes_n$. We can deduce that $\mathbb{Z}_n^* = \{ a \in \mathbb{Z}_n : a \text{ is prime with } n \}$ (see the case of

$n = 4$ above). This is a consequence of the Bachet–Bezout identity, which ensures that, if a is prime with n, there exists an integer b in $\mathbb{Z}_n$ and a natural number c such that $ab = nc + 1$. The element b is then the inverse of a for $\otimes_n$ and vice versa). Both are therefore in $\mathbb{Z}_n^*$. We then check, as before, that we can endow $\mathbb{Z}_n^*$ with the multiplication $\otimes_n$ (according to the same definition) and that, for this operation, every element of $\mathbb{Z}_n^*$ has an inverse, making this set a commutative (multiplicative) group.

In the case of a p-primary number, we therefore consider to determine $\mathbb{Z}_n^*$ only the (non-zero) integers of $\{0, \ldots, p^r - 1\}$ which are prime with $n = p^r$, which are, in this case, those not divisible by p. It is easy to assume that there are exactly $p^r - p^{r-1} = p^{r-1}(p-1)$.

When p is different from 2, as in the case where n is itself a prime number other than 2[9], $\mathbb{Z}_n^*$, is cyclic, in the sense that it possesses generator elements, that is to say, elements a of $\mathbb{Z}_n^*$ whose successive $\otimes_n$-powers go through all $\mathbb{Z}_n^*$ (see Perrin 1996, Proposition 3, Chapter I). The paths of these generators – of length $p^{r-1}(p-1)$ – are also very erratic, at least for some values of a.

When $p = 2$, things get even more complicated. Indeed, when $n = 2^r$, if $\mathbb{Z}_n^*$ always counts $2^{r-1}(2 - 1) = 2^{r-1}$ elements and can be safely endowed with the $\otimes_n$-multiplication, it can no longer be generated by the successive powers of some of its elements, as previously seen (see again Perrin 1996, Proposition 3, Chapter I). However, its structure remains relatively simple[10] and the successive $\otimes_n$-powers of some of its elements can exhaust 2^{r-2} distinct elements of $\mathbb{Z}_n^*$, that is to say, one in two, while still having the sought-after erratic character. We will come back to this briefly, because this loss of a simple factor 2 has been found, over time, to be largely compensated for by the virtues of the powers of 2 as related to the binary arithmetic of computers. The designers of vector-supercomputers at the end of the 20th century were able to take advantage of this.

2.5.2. From erratic arithmetic to algorithmic randomness

After these few steps, let us come back to the prime numbers themselves, from where it all started. From the erratic to the random, there is only one, admittedly risky, step (and vice versa), but it is one which the pioneers of the Monte Carlo simulation took, not lightly, as one might have imagined, but laboriously and after much discussion, hesitation and a mixture of unsuccessful attempts.

---

9 When $n = 2$, $\mathbb{Z}_2^* = \{1\}$ is still cyclic, because it is generated by 1, but this trivial case is of little interest.

10 In fact, in technical terms, $(\mathbb{Z}_n^*, \otimes_n)$ is "isomorphic" as a group to $\mathbb{Z}_2 \times \mathbb{Z}_{2^{r-2}}$ endowed with the addition component by component, that is to say that there is a bijection between these two sets satisfying that $\varphi(a \otimes_n b) = \varphi(a) \oplus_{2,2^{r-2}} \varphi(b)$.

The guiding intuition, in modern terms, is to go to the scale in the phenomenon that we have just observed by taking, for n, not a prime number of one or even two digits, but a very large prime number, for example, chosen from the Mersenne series of numbers of the form $2^p - 1$, where p is itself prime, which deemed to have many prime numbers within it[11]. By "large" we mean really, very large.

It is to the famous number theorist D.H. Lehmer in 1949 (Payne *et al.* 1969) that we owe the idea of how to exploit the above observation and therefore, in particular, to scale it so as to build a pseudo-random number generator, combining algorithmic efficiency and simulation quality (at least for that period in time). His initial proposition, according to (Park and Miller 1988), considers the above recursion (*) with $n = 2^{31} - 1 = 2,147,483,647$, the eighth Mersenne prime number, first discovered by Euler in 1750[12], and $a = 14^{29}$. This second choice may be puzzling with regards to the computational capacities of the era[13]. This class of generators is known today as *Lehmer Random Number Generators* (*Lehmer RNG*).

At the dawn of the 1950s, all of this was in keeping with the times. Mathematically, number theorists have always been fascinated by both this quest for records and the endless challenge it represents. Yet, this quest had stagnated for more than fifty years: the last discovery – the twelfth Mersenne prime number, $2^{127} - 1$ – was by Edouard Lucas back in 1876. We were reaching the limits of the calculation by hand approach. It was not until three quarters of a century later, with the appearance of the computer, that we were able to restart this race, which had up until then seemed a little futile.

Indeed, the next record was once again beaten by a Mersenne prime number, the thirteenth, $2^{251} - 1$, which was only discovered in 1952, but this time with the crucial "support" of a computer program designed by R. Robinson and running on a new computer called *SWAC*, one of the first successors to *ENIAC*. Progress after this was meteoric, this time keeping in mind the design for pseudo-random number generator applications as one of the main driving motivators. Let us quote B. Tuckerman (IBM) who, in 1971, established that $n = 2^{19937} - 1 \simeq 10^{1994}$ is the twenty-fourth Mersenne prime number, setting a new record. In those years, the record was beaten with metronomic regularity, often, but not always with Mersenne prime numbers.

---

11 The sequence begins as follows: $2^2 - 1 = 3, 2^3 - 1 = 7, 2^5 - 1 = 31, 2^7 - 1 = 127$ are prime, $2^{11} - 1 = 2,047 = 23 \times 89$ is not, but $2^{13} - 1 = 8,191$.

12 This Mersenne number also has the distinction of being one of the four known *Double Mersenne Primes* in the sense that 31 is, in fact, also a Mersenne number.

13 The justification given by Lehmer is as follows: the cardinal $2^{31} - 2$ of $\mathbb{Z}_n^*$ breaks down into prime numbers in the form of $2^{31} - 2 = 2 \times 3^2 \times 7 \times 11 \times 31 \times 151 \times 331$. Now 14 is a generator of $\mathbb{Z}_{2^{31}-1}^*$. In addition, we easily verify via Bezout's identity, which raises 14 to a prime power of r with $2^{31} - 2$, here $r = 29$, creating a new generator $\mathbb{Z}_{2^{31}-1}^*$ introducing more "randomness" in the number sequence. The remainder $14^{29} > 8^{29} = 2^{63} > 2^{31} - 1$ (by a lot!); poses some particular arithmetico-computational difficulties.

Why then the attachment to this particular record-breaker? Quite simply because this number is currently still at work in the most famous of the pseudo-random number generators, the *Mersenne Twister 623* (see Matsumata and Nishimura 1998), even though it does not directly fit into the elementary framework of congruent generators we are currently describing. Today, the largest prime numbers are through the roof[14], however, for the simulation of pseudo-random number sequences, such a size would be more than sufficient, since changing the base of the prime number would require one to re-start the design and validation of the generator, with no guarantee of any significant improvement. In the meantime, other fields of applied mathematics have taken over and are eagerly searching for these very large prime numbers, without ever finding satisfaction, it seems. This is particularly true for the case of cryptography. Besides, this race – or rather, this escalation – which we know to be endless, feeds on itself out of a universal fascination for prime numbers that is enough without needing to worry too much on its possible applications.

In order to implement the project of making a generator from the previous idea, we still need to find a primitive root! In this case, we have to find a primitive root $a \in \mathbb{Z}_n^*$ (necessarily different from 1) whose successive $\otimes_n$-powers generate (run through) all Z_n, which is no small matter, especially when n becomes large. Moreover, and this is the crucial point, the powers of this number a must run through $\mathbb{Z}_n$ as erratically as possible, such that the sequence of numbers:

$$\frac{a^{\otimes_n k}}{n}, \quad k = 0, \ldots, n-2$$

where the division $\div$ denotes the usual division between real numbers, and can be considered as realizations (admittedly approximate, but statistically acceptable) for a sequence of random numbers distributed in the interval $[0, 1]$. To vary the pleasures and further one's enjoyment from a "walk" through the generator, it is customary to let the user choose a *seed* $c \in \mathbb{Z}_n^*$ (i.e. $c \neq 0$!), and then consider the following:

$$(\star) \qquad \frac{c}{n}, \frac{c \otimes_n a}{n}, \ldots, \frac{c \otimes_n a^{\otimes_n k}}{n}, \ldots, \frac{c \otimes_n a^{\otimes_n (n-2)}}{n}$$

This operation allows one to exhaust the initial sequence, starting from the term $\frac{a^{\otimes_n k_c}}{n}$, where $a^{\otimes_n k_c} = c$ (the exponent k_c exists since the $\otimes_n$-powers of a exhaust all $\mathbb{Z}_n^*$). If we take for granted that a good choice of a will indeed generate a satisfactory imitation of random numbers uniformly distributed in $[0, 1]$, why then favor this approach? The reason is purely algorithmic: the sequence $c \otimes_n a^{\otimes_n k}$,

---

14 The largest known Mersenne prime, and probably the fifty-first number, is $2^{82589933} - 1$. It is made up of $24,862,048$ numbers in base 10 and was discovered in December 2018 by P. Laroche through the GIMPS (Great Internet Mersenne Prime Search) Project, the base of all recent discoveries of the largest Mersenne prime numbers. It is the largest prime number known at the time of writing.

$k = 0, \ldots, n - 2$, above is obtained by successive iterations of the transformation from $\mathbb{Z}_n^*$:

$$k \longmapsto k \otimes_n a$$

from the seed c. It is therefore sufficient to code the multiplication $\otimes_n$ in low level language so as to obtain a quick simulation algorithm for $\frac{c \otimes_n a^{\otimes n k}}{n}$. This is the major asset of Linear Congruent Generators. A variant consists of adding a kind of *drift* b to the transformation, i.e. to consider:

$$k \longmapsto \left(k \otimes_n a \right) \oplus_n b$$

without this significantly changing the nature of the generator's performance. However, this complicates things, because $\mathbb{Z}_n^*$ is not stable by $\oplus_n$-addition (if a and b are in $\mathbb{Z}_n^*$ their sum cannot be: therefore, in $\mathbb{Z}_5$, 3, 2 there are elements of $\mathbb{Z}_5^*$, but $3 \oplus_5 2 = 0$). The successive iterations of this transformation resulting from a seed c, supposedly different from the single forbidden seed c_0, will, however, go through all of $\mathbb{Z}_n$, except one of its elements; which does not depend on the seed c chosen[15].

This process, by virtue of its speed and its relative simplicity, defeats the abstract definitions (of both logicians and theoretical computer scientists) on the notion of random number sequences. However, there is no guarantee, other than empirical, authenticating the randomness of a sequence produced by a congruent generator. What a theoretical approach to develop, since, if one were to iterate the above transformations beyond the limits indicated, one would see a return of the first values in the sequence. The reason for this is simple: if we extend the iterations indefinitely, the sequence obtained is periodic! This inherent flaw for this type of generator has led to the recommendation of using only 10 to 15 % of a given simulation. For modern generators (see below), this is hardly a constraint anymore, since their period allows for simulations with several billion scenarios in high dimensions (several hundred).

To validate such a generator, various statistical tests have been developed, often inspired by probabilistic properties and established for the generic realization of an abstract random number sequence[16]: the maximum length of an increasing series, independence by packets, etc. Various toolboxes have been developed, notably under the impetus of G. Marsaglia, to test and challenge all generators aspiring to deserve this "title".

---

15 When $b \neq 0$, the forbidden seed is $c_0 = n - (a - 1)^{\otimes n(-1)} \otimes_n b$ and the unvisited value is $n - (a - 1)^{\otimes n(-1)} \otimes_n b$: we check that if $b = 0$, the forbidden seed is worth $n \ldots$ therefore 0 "modulo" n since $n = 1 \times n + 0$.

16 In probabilistic language: the generic realization of a sequence of random variables, identically independent and distributed uniformly over the interval $[0, 1]$, but which is, let us agree, hardly enlightening for the non-specialist.

Today, the principle described above has been made both more complicated and robust by the use of several generators in parallel: a combination of generators, random picking in a generator via an auxiliary congruent generator, etc.

Congruential generators are not the only methods used for the generation of pseudo-random numbers (as is the case with permutations from registers which are based on both the architecture of arithmetic processors and the theory of finite fields), nor the manual methods, which, far from having had their last say, are today experiencing a renewed interest, as they are supposed to generate "authentic" random numbers.

2.5.3. *And, the winner is...* Mersenne Twister 623

Among the most used generators of today, many are still of the linear congruential type, like the *Lehmer RNG*, often in their "affine" variant ($b \neq 0$). The best known is undoubtedly the implementation by Park and Miller (1988), who, in 1988, proposed, once again with $n = 2^{31} - 1$, replacing the huge and inconvenient 14^{29}, put forward by Lehmer himself, with the more reasonable $a = 7^5 = 16, 807$.

As for the *RANF* instruction (for "RANdom Function") of Seymour Cray's vector supercomputers in the 1980s, which definitely strayed from historical paths by discarding their evolution over $n = 2^{48}$ and $a = 44, 485, 709, 377, 909$, so as to obtain a period generator $2^{48-2} = 2^{46}$, better suited to massively vectorized computations (the precursor to parallel computation)[17].

Today, the most prevalent and commonly used generator of pseudo-random numbers is the *Mersenne Twister 623*, developed by two researchers from the University of Hiroshima (in a bizarre twist of historical irony), based on the Mersenne prime number $n = 2^{19937} - 1$ mentioned above. Let us make it immediately clear that this is not, strictly speaking, a simple congruential generator as described above, if only due to its vectorial nature and the use of XOR ("or" exclusive) logical functions in its update structure to pass from one number to the next. Nevertheless, it retains a recurrent structure whose precise description goes far beyond the scope of this paper.

It successfully passes the most rigorous statistical validation tests up to the 623-dimension, at the cost of a few improvements. This means that if, in practice, one were to regroup the successive numbers into packet lengths of 623, the vectors obtained would successfully pass the various statistical independence tests in dimension 623, for toolboxes dedicated to a significant fraction of the generator. Despite the ever-increasing size of current MC simulations, in neutronics, statistical

---

17 It is a simple exercise to show that if $2^n - 1$ is prime then n must be as well.

physics or finance, the risk of it becoming exhausted is still a long way off: a MC simulation of 10 billion scenarios from the *Mersenne Twister 623* generator, consuming 500 numbers per scenario, will only consume a completely negligible fraction, in this case $2/10^{1982}$.

Of course the story does not end there. With with the appearance of parallel computing, in particular on Graphical Processing Units (GPUs), the need for generators that can produce very large random vectors extremely quickly has become imperative. Indeed, each thread (unit responsible for a stand-alone micro-task) on the board consumes its own random numbers, and up to more than one million threads (a set of micro-calculations) can be executed simultaneously on such a card. Therefore, the recently developed *Philox* generator, which also privileges 2-primary numbers, in this case $2^{256} \simeq 10^{77}$, still because of binary arithmetic computation units, seems far better suited to parallel computation than the *Mersenne Twister 623*: it is able to efficiently initialize a million threads simultaneously – which can be similar to so many Monte Carlo simulations – a thousand times faster than the *Mersenne Twister 623*, improving the expected performance of the GPU simulation. This *Philox* generator is not really a *Lehmer RNG*-type congruential generator, because the pseudo-numbers generated randomizations are not obtained by a simple division of integers calculated by a multiplicative induction, but rather the result of a much more complex transformation from a recursion of the results (see, for example, Salmon *et al.* 2011).

And so, in answer to the question that also serves as the title of this appendix, we may say "yes", and even if this is as it should be, it is fortunately a little more complicated than that.

Thanks to Vincent Lemaire for his help in writing this appendix, and my thanks to the team at ISTE for the additional reference about Galileo's problem.

2.6. References

Bachelier, L. (1900). Théorie de la spéculation. *Ann. Sci. École Norm. Sup.*, 3(17), 21–86.

Bernoulli, J. (1713). *Ars conjectandi*. Thurneysen Brothers, Basel.

Black, F. and Scholes, M. (1973). The pricing of options and corporate liabilities. *Journal of Political Economy*, 81(3), 637–654.

Boyle, P.P. (1977). Options: A Monte Carlo approach. *Journal of Financial Economics*, 4(3), 323–338.

Borel, E. (1920). *Le Hasard*. Félix Alcan, Paris.

Buffon, G.L., Comte de (1733). Mémoire sur le jeu du Franc Carreau. Presented, Académie Royale des Sciences, Paris.

Hairer, M. (2014). A theory of regularity structures. *Invent. Math.*, 198(2), 269–504.

Harrison, J.M. and Kreps, D.M. (1979). Martingales and arbitrage in multiperiod securities markets. *J. Econom. Theory*, 20, 381–408.

Harrison, J.M. and Pliska, S. (1981). Martingales and stochastic integrals in the theory of continuous trading. *Stochastic Processes and their Applications*, 11, 215–260.

Hombas, V.C. (2004). Generalizing Galileo's passedix game. *Interstat*, 1 [Online]. Available at: http://interstat.statjournals.net/YEAR/2004/articles/0401001.pdf.

Kolmogorov, A.N. (1933). *Grundbegriffe der Wahrscheinlichkeitsrechnung*, Foundations of the Theory of Probability. Julius Springer, Berlin.

Lévy, P. (1940). Sur certains processus stochastiques homogènes. *Compositio Mathematica*, 7, 283–339.

Lévy, P. (1970). *Quelques aspects de la pensée d'un mathématicien*. Blanchard, Paris.

Matsumata, M. and Nishimura, T. (1998). Mersenne twister: A 623-dimensionally equidistributed uniform pseudorandom number generator. *ACM Trans. in Modeling and Computer Simulations*, 8(1), 3–30.

Metroplis, N. and Ulam, S. (1949). The Monte Carlo method. *J. Amer. Statist. Assoc.*, 44, 335–341.

Pagès, G. and Bouzitat, C. (2003). *En passant par hasard*, 3rd edition. Vuibert, Paris.

Park, S.K. and Miller, K.W. (1988). Random number generators: Good ones are hard to find. *Communications of the ACM*, 31(10), 1192–1201.

Pascal, B. (1897). *Pensées*, Brunschvicg edition. Hachette, Paris.

Payne, W.H., Rabung, J.R., and Bogyo, T.P. (1969). Coding the Lehmer pseudo-random number generator. *Communications of the ACM*, 12(2), 85–86.

Perrin, D. (1996). *Cours d'algèbre, maths AGREG*. Ellipses, Paris.

Robbins, H. and Monro, S. (1951). A stochastic approximation method. *Ann. Math. Stat.*, 22, 400–407.

Salmon, J.K., Moraes, M.A., Dror, R.O., and Shaw, D.E. (2011). Parallel random numbers: As easy as 1, 2, 3. *SC11*, November 12–18, Seattle, USA.

3

Chance in a Few Languages

To Jean-Michel Mehl
and in memory of Dominique Lacroix

In what manner, with which words, through what expressions, do we convey the notion of *chance*? It is this very question that motivates the lines that follow. We will go quickly, taking the word or words that can mean chance in a few languages: classical Sanskrit, Persian and Arabic, ancient Greek and Russian (very quickly for the latter), Latin, French (with three words for Italian and Spanish), finally English (with two words for German). Such an investigation would be incomplete if we did not also take into account the social facts, such as divination and the game of dice, in short if we did not take heed of the designation that these words have. I am not confusing "meaning" and "designation"; the first evokes the etymology of a word, its grammatical and lexical history; the second its historical, practical and precise usages, which allow a word to designate, point out in some way, a person or an object, an action or an idea.

A few preliminary remarks are in order. Certain specific meanings for "chance" were eliminated from the French survey: the meaning of *risque* ("risk"), as in: "to put something at stake, or run the possibility of"; the meaning of *but imprécis* ("undefined goal") as in *à tout hasard* ("by any chance"). The expression *ce n'est pas un hasard* ("it is not by accident") is a way to support a trivial demonstration by means of giving evidence of something.

I have therefore kept the following linguistic senses and situations of the French:

– *cause* ("cause"), objectively judged as unnecessary and unpredictable, of events which may however be subjectively perceived as intentional;

Chapter written by Clarisse HERRENSCHMIDT.

– *concours de circonstances* ("culmination of circumstances"), the outcome of unexpected or inexplicable events (products of chance), synonymous with *coincidence*; fate, fortune, fortuitous events, etc.;

– *jeu de hasard* ("game of chance");

– *par hasard* ("by chance"), which according to the digital French Language Treasury[1] means: in phrases, to find oneself, to be, to happen, "by chance", "by accident", etc.

3.1. Classical Sanskrit

If this chapter begins with Sanskrit, it is due to the fact that this great language of civilization has a great body of work which has enjoyed immense reflection, and not because Sanskrit is said to be *the* origin language of the Indo-European languages – which will be the overwhelming focus here.

Classical Sanskrit is not the oldest language in the Indian Brahmanic tradition; that would be Vedic Sanskrit, the language of the Veda. Classical Sanskrit slowly began to emerge from the 5th century BCE, solidifying around the 2nd century CE. The Indian languages that derive from it build upon and grow according to the precepts of Indian grammar – the oldest linguistic group in the world.

In the tradition expressed by classical Sanskrit, there are several words, almost all compounds, for *luck* and *chance*. Let us just observe their composition. A number of these are formed from the preverb *sam* ("together") and a nominal derivative of a verbal root. This can be *bhû* ("to be, to become") from where *sam-bhâvanâ* ("the fact of assembling, to agree"), or *yuj* ("to bind") become *sam-yoga* ("bond, connection, astronomical conjunction, marriage") which is the origin of the modern Hindi word *samyôg* ("luck, chance, combination, connection, good fortune"); or even *gam* ("to go"): *sam-gati* ("meeting, fact of agreeing"), *sam-gatyâ* ("if applicable"). Other words were formed with *daiva* ("divine"), derived from *deva* ("god"), which gave rise to *daiva-yoga* ("that which is bound to/created by a god") and *daivatas* ("fatally, by fate, by luck, by chance"). Finally, others were formed with the (neutral) pronoun *yad* ("which, that") and a derivative of a root, for example *ricch* ("to occur") such as *yad-ricchâ* ("accident, chance") and *yad-ricchatas* ("by chance").

We will deduce that in a few words for *chance* in classical Sanskrit, there can be a divine action, or a prior action, or for example that of "connecting", whose driving force is unknown, or, finally, nothing at all, neither a divine action, a prior action or a driving force, as in our last two examples.

---

1 *Trésor de la langue française informatisé* or TLFi.

Yet, I did not come across an expression for which chance was invoked thanks to the game of dice, despite the tales whose backdrop depict kings and nobility at play – the most famous being a passage from the *Mahabharata*, in which King Yudhishthira loses all of his estates (and his wife) to a demon in a game of dice. While it is clear that this investigation is incomplete, it is also possible that the metaphor of chance in India is of a different nature. In fact, the word for "dice" is *pâça* ("bond, rope, chain, lace, trap, dice") to which is added the root *yuj* ("to bind") and used in the formation of words evoking chance, which allows one to hypothesize that the founding metaphor of the expression of *chance* in India would be that of bondage and attachment, and even, convenience.

3.2. Persian and Arabic

The following is based on a written interview with Leïli Anvar, a teacher at the *Institut National des Langues et Civilisations Orientales*[2] and an excellent Persian source for our discussion on *chance*, reproduced here with her permission.

> This is an extremely interesting question that I return to regularly with my students. I would say, to use a short and sarcastic phrase: in Persian, "chance does not exist"! Explanation: the word for it is of Arabic origin (I don't know how we Iranians expressed *chance* before the Arab conquest – or if this was said – in the older languages of pre-Islamic Iran, in Middle Persian for example, since all languages do not have the notion of chance, as you know): *ettefâq* (اتفاق), which comes from the root w-f-q, referring to the notion of agreement, compliance, conformity, accordance. The root verb of *ettefâq* is *ittafaqa* – in Arabic – which means "to fit, to coincide, to accord, to correspond, to agree". So, and this is the fascinating metaphysical aspect: *ettefâq* by which we designate *chance* and *coincidence* also means the "thing which coincides, which is exactly in tune with the situation". The underlying reason for the choice of this word to mean *chance*, is in my opinion, the idea that there is only "apparent chance". In reality what seems to happen haphazardly, by luck, without a specific reason or cause, is something which coincides perfectly with the situation, seamlessly fits with it, and therefore: the word that expresses *chance* in Persian and in Arabic, simultaneously says that, chance both exists and does not exist. Or at least, that chance is a necessity, which

2 A French research institute teaching languages that span Central Europe, Africa, Asia, America and Oceania. It is often informally called Langues O' or, more recently, goes by the acronym INALCO.

amounts to the same thing! Furthermore, the great Persian dictionary *Sokhan* gives as the fifth meaning of the word *ettefâq*: destiny.

The other thing this little impromptu wants to demonstrate is that there is no better approach to metaphysics than etymology. Words tell us so much about reality as it is experienced in the body of thinking! In short, what better way is there than language and words to nourish our existential meditation? The champions in this area are the Jews with their truly dizzying approach to Hebrew. I don't know of a better model for thinking than the Talmudic model, and Ibn Arabi does the same with Arabic, but I believe, with all due respect to the Sufis, that he was in fact strongly influenced by the Jewish thought prevalent in his native Andalusia.

3.3. Ancient Greek

In ancient Greek, chance was called *hè tychè*, the original meaning of which was "act of a god" and which as a result became a proper name: the goddess Tyche is often depicted holding a cornucopia from which flows agricultural goods or coins. To show the unstable character of human situations, the wheel of fortune sometimes appears, at other times Tyche is standing on a circular shape, and of course, is blindfolded. The semantic developments of *tychè* consisted of: "the act of a human being", then, as a cause beyond human control, "necessity, destiny, fortune, fate, chance". Very difficult to translate here, Plato said that *tychè* carries events brought about by the action of a non-human, "from the invisible to the invisible".

"By chance" was expressed by *en tychè*. The syntagm *syn tychai* ("with favorable divine action, by chance"), already used in classical Greek, was later generalized by Byzantine Greek.

Another way of saying *chance* is the neutral substantive *to automaton* ("what dies of its own accord, what acts due to its own movement, the spontaneous, the natural"), opposed to *hè technè* ("art, artifice, the means to implement"); *ek automatou* and the adverb *automatôs* could be translated as "without external action, by chance".

A sort of personification of *chance* can be seen elsewhere in ancient Greek other than the divinization of *hè tychê*; for instance, to say "to occur/to happen by chance, something happens to someone", the Greeks used the verb *tygchavô*, from the same root as *hè tychè*, conjugated in both singular and plural, whose grammatical subject appears under the form of a participle, for example: *tote etygchanon eisiôn*, "so I entered by chance", actually means, word for word, "I happened to be entering then".

It is the subject of precise and spatiotemporal action that implies chance, similar to the grammatical structure of the verb *to happen* in English.

Greek has also known impersonal forms: (*hôs*) *etyche* ("as it happened, by chance"). This same verb allows the word *tychon* to mean "maybe".

In addition, the die was said to be either *ho kybos* ("cube, die with a point on each of its six faces"), or *astragalos* ("the knucklebones"), in the plural *astragaloi* ("to throw the bones") – it was made from a young sheep's tarsus, the sides of which were dotted – in the plural because they played with three, four or five bones; a "game of chance" known as *hè kybeia*.

The use of *kybos* denotes that it designated *chance*: *ergon in kybois Arês krineî*, "Ares will decide the issue with his dice" (Aeschylus *Seven Against Thebes*, v. 414). Or again: *aei gar eu piptousiv hoi Dios kyboi* (fragment 895 of Sophocles), "the dice of Zeus always fall luckily", that is to say "the will of Zeus is not exposed to any chance". On itself *kybeuô* means "playing dice".

3.4. Russian

The Russian word случай (pronounce *sloutchaï*), meaning "chance", comes directly from the Byzantine Greek syntagm *syn tychai* ("thanks to divine *tychè*"), as the conceptual Russian language bears a strong imprint of Byzantine Greek, the language of the first orthodoxy; in Russian, the two Greek words *syn* and *tychai* have been merged into one. In addition, Russian and languages such as Polish and Czech borrowed from the French words *chance* ("luck") and *risque* ("risk") (the latter is in fact derived from Italian). шанс and риск, in Russian, also express this idea of *chance*.

3.5. Latin

Let us take two examples from Cicero; *sed haec ut fors tulerit* ("but that as fortune wills") (letter to Atticus, 7, 14, 3); *sed haec fors viderit* ("but that which we must leave to chance") (id. 14, 13, 3). Chance or luck, in Latin, *fors*, is something that produces an outcome that we cannot presently know. The word belongs to the verbal root *fero, ferre, tuli, latum* ("to bear, carry, suffer, endure"). The Latin substantive *fors* has only two of the six cases of the Latin declension; a nominative where this word is the subject of a verb, "to bear" or "to be" (*fors fuit*, "chance of"), and a singular ablative: *forte* which has become an adverb, "by chance". Why is the idea of "carrying, bringing" associated with the notion of *chance*? It seems to stem from a very practical aspect of divination for which the Romans had frequent use; divination could be done according to various methods, including the following:

various objects such as inscribed wooden tablets were enclosed in an urn, the innocent hand of a child would take one out, hold it up to the light and read its notes and these signs were the answer to the question being asked – for example: whom to assign to a particular public office.

Fortuna was the proper name for the goddess of fortune, often evoked by the syntagm *fors Fortuna*, identified in tandem with the Greek goddess Tyche and represented in the same way. This is Good Fortune, that of the Roman people or that of Caesar; as a common noun *fortuna* means "luck, contentment (without qualifier), social condition".

There are several words which mean "maybe", some of which formed from *fors* (*forsan, forsit, forsitan*) and from *forte* (*fortasse, fortassean, fortassis*). The latter evokes a possibility and only *forte* ("by chance") evokes the notion of a result, realized or possible, from a blind draw.

We all know the sentence attributed to Caesar by Suetonius: *Alea iacta est* (*Life of the Twelve Caesars*, Caesar 32), "let the lot be cast".

Dice were called *alea* and "playing dice" were *alea ludere*; *ad talos*/*ad tesseras se conferre* means "to indulge in the game of bones/dice", *aleam exercere* means "to play games of chance"; in Cicero we already find *aleator* ("the dice player") and the adjective *aleatorius* ("which concerns the game of dice") – the origin of the French word *aléatoire* which we will come back to later. Randomness in terms of the "game of dice, game of chance" is made clear by Plautus (Curculio, 355): *talos poscit sibi in manum, provocat me in aleam, ut ego ludam* ("he calls for dice, and challenges me to the gamble with him for stakes"); the meaning of "chance, luck, risk" is found in Cicero (*De Divinatione*, 2, 36): *non perspicitis aleam quandam esse in hostiis deligendis?*, "do you Stoics fail to see in choosing the victim it is almost like a throw of the dice?"

The Latin *talus* refers to a knucklebone with four marked sides and *tessera* the cubic die marked by six sides; the dice or the bones were thrown (in Latin, *jacere*) after which they fall (in Latin, *cadere*). In this regard, let us also quote Cicero: *talus erit jactus ut cadat rectus*, "a bone thrown so that it lands straight" (*De Finibus Bonorum et Malorum*; "On the ends of good and evil" III, 54).

A very common word, *casus* ("fall, fortuitous event, accident, circumstance") was used in the ablative *casû* by Cicero and others to mean "by chance"; this word, derived from the verb *cadere* ("to fall, to die, to arrive, to happen") can be found in Italian in the form *caso*, in French as *cas* and *occasion* and in English as "case" and "occasion".

3.6. French

The French word for chance, *hasard*, is reputed to be of Arabic origin, as the TLFi states: "most of the dictionaries consulted say the word *hasard* originates from the name for a game of dice in the Middle Ages, with the term *hasard* being the output of a number of points favorable to the player. Borrowed from the popular Arabic *az-zahr* 'playing dice' by way of the Spanish *azar*, the term designates a 'move unfavorable to the game of dice, a type of dice game'" (1283). It seems that the meaning of "playing dice" for this Arabic word arrived late and does not appear in classical dictionaries; another proposition was that "the Arabic *yasara* 'to play dice', or *yasar* 'group of dice players', derived from the same verb attested to in classical Arabic". Among the modern Arabic names for die is the word *zahr*, which is clearly from this same root. In short, although the Arab origin of *hasard* ("chance") is certain, it is not absolutely clear.

In 14th-century France *par cas* ("by case") meant *par hasard* ("by chance"), although this usage has since disappeared, while the Italian equivalent *per caso* ("by chance") survives to this day.

The two statistically closest words to chance according to the Dictionary of Synonyms of the University of Caen (CRISCO) are *chance* ("luck") and *aléa* ("random"). Let us quote the entry for *chance* from the TLFi: (1) *Circa* 1175 *chaance* "way (generally favorable) in which an event can change" (Benoit de Sainte-Maure, *Chronique des ducs de Normandie*, composed around 1175); hence 1762 (Jean-Jacques Rousseau, *Du contrat social*, III, 6): "[...] they put almost all the *chances* against themselves"); (2) 1200 *caanche*, "fall of the dice" (Jean Bodel, *Le Jeu de Saint Nicolas*, composed at the beginning of the 13th century). This is the substantivation of the Latin *cadentia*, plural neutral present participle of *cadere* ("to fall") which was also used in classical Latin in the lexicon of the game when speaking of knucklebones (Cicero, *De finibus bonorum et malorum*, 3, 54).

The Latin verb *cadere* gave the Old French *cheoir*. *Cheance*/*chance* is first "the fact of letting the dice fall, the roll of the dice", but the medieval game which was entitled *Deux dés à trois chances* indicates that *chance* (Medieval French) also designated the "turn, the order in which each player takes their turn to roll"; obviously, *chance* (id.) designated the outcome of the point counting, that is the correct result; thus, when at the foot of the cross of Christ the Roman soldiers play dice declared "the best *chance* (id.) wins", meaning "the best outcome wins".

The French word *aléa* comes from the Latin *alea* ("dice, game of dice"); let us again reference the TLFi: 19th century (vocabulary in) banking ("chance of gain or of loss in the course of business or speculation") (Circular advertisement inserted in the financial newspapers by the administrators of the railroad company in charge of the route between Orléans and Rouen in Littré *Addenda*, 1872: a fixed price contract was made with responsible contractors, for the completion of the line, at the rate of 150,000 francs per kilometer, including the interest and overheads encountered during construction, and which protected the company from any kind of bad luck); hence, in 1873 the more general use of the term *chance incertaine* meant "uncertain luck".

To speak of *hasard* ("chance"), we need adjectives. *Hasardeux* ("risky") is not suitable, because it can have multiple synonyms: uncertain, flimsy, daring, perilous, audacious, in bad taste, etc. It is therefore *aléatoire* ("random") that is typically used.

However, a competitor to *random* has since entered onto the international scene, in mathematical language: *stochastic*. However, I was unable to find the original text, in English or otherwise, in which it first appeared. We note that substantive stochastic also exists. TLFi gives the following definitions:

> I. Adjective. A. *Epistemology*. Which depends, which results from chance. B. *Mathematics, statistics*. 1. Which comes under the domain of randomness, the calculus of probabilities. 2. Which uses probability theory. C. *Computer science*. Stochastic calculator. II. Feminine noun, *mathematics, statistics*. A branch of mathematics that deals with the use of statistical data by calculating probabilities. (Dict. 20th century)

The etymology of *stochastic* does not present any difficulty: it is borrowed from the Greek, in the adjectival form: *stochastikos* ("skillful in aiming"), formed of the verb *stochazomai* ("to aim, to shoot, to try to distinguish, guess, calculate, conjuncture"), or on the nominal form: *hê stochastikê* ("the process by conjecture"); we read in Plato *hê stochastikê technê*, "the art of proceeding by conjecture", and in the fragments of the ancient Stoics: *stochastikai epistêmai*, "the sciences of conjecture". The family of words whose base is *stochos* ("the goal, the target") has also produced the word *stochastês* ("the diviner").

The Cerisy conference brought to my ear for the first time the word *stochasticité* ("stochasticity"). As the researcher who used it told me, this word exists in English; in response to my question: "why not use the word *chance*?", he expressed the idea, then and there, by expressing himself freely, of reserving *chance* for the concept of chance in a deterministic environment like that of physics and *stochasticity* for the concept of chance in a non-deterministic environment, like that of biology. This constitutes a significant contribution to language and thought – we will come back to this.

Italian borrowed from the French the word *hasard* by Italianizing it: *azzardo*, as in *gioco d'azzardo* ("game of chance"); its oldest word for the notion is *caso* ("unforeseeable cause"), which, as seen above, comes from the Latin and also evokes the object: dice; "at random" is *a caso*; "on the off chance" is *a ogni caso*; "by chance" *per caso*, finally "coincidentally", ironically, *guarda caso*.

Spanish plays on two registers: that of Latin *cado* with *casualidad* ("occasion") and *acaso* ("by chance") and borrows from the Arabic *azar*.

In German, "chance" is called *der Zufall* and "by chance" is *zufällig*; the word comes from the root *fallen* ("to fall, to die, to throw itself, to happen"); *Fall* ("the case"), *auf jeden Fall* ("in any case"). We therefore find with the verb *fallen* ("to fall") and the substantive *der Fall* ("the case"), *der Zufall* ("the chance") the same situation as that of Latin (*cadere, casus, casû*), but I do not know if *der Zufall* implies the game of dice; *alea iacta est* translates as *die Würfel sind gefallen*.

3.7. English

English has a vocabulary of dual origin, Saxon and French. The words available are *chance, luck* and *random* – the latter being the scientific concept.

Chance was borrowed from the 13th-century French "*jeu de hasard*" or game of chance, and refers to the game of dice; *main chance* acquired the mathematical sense of probability, especially in gambling, as early as 1778. *Luck* is of dubious origin and does not appear until the 16th century; "maybe a game term" they say, perhaps affiliated with German *(g)lück*, which means "happiness".

The scientific concept of chance is said to be *random*, or *at random*, and can be both the noun and the adjective. Its origin consists of the old French *randon* ("briskly, impetuosity, precipitation, disorder"). The word was already rare in classical French, but we find it in Jean de La Fontaine (*Poésies mêlées* XXVII):

L'hiver survint avec grande furie.
Winter came with great fury.
Monceaux de neige et grands randons de pluie.
Snow drifts and great squalls of rains.

The origin of the French *randon* happens to be the verb from the Old Frankish *randir*, "to run, to move quickly", from Old High German, and which gave the German verbs *rennen* and the English *run* (the same meaning in both languages). The French word *randonnée* ("to hike") is a nominal derivative from the generalization of *–ée* of the old French *randon*, a derivative which has since lost the connotation of "running" as we know it. The English *random* and the French *randonnée* are not only of the same family, but are extremely close, even though they no longer mean the same thing – chance versus an excursion!

The Arabic *zahr* passed from Spanish into French as early as the 13th century. The English verb *to hazard*, in the sense of "playing dice", with its different *chances*, and the noun *hazard* "game of dice" were borrowed from the French. The adjective *hazardous*, meaning "dependent on chance" (Oxford Dictionary of Current English), comes from the French *hasardeux*, and occupies the same semantic field as the game of dice. The etymology of *haphazard* ("at random, hit and miss") is explained by the agglutination of *hap* + *hazard*, with *hap* coming from the same family as *happy*.

From this point forward we shall focus on the game of dice.

3.8. Dice, chance and the symbolic world

The essential sources used by the historian J.M. Mehl for his admirable work are the Letters of Remission (Mehl 1990), a Chancellery document granting pardon to someone who was accused of a crime and who, thanks to this letter, was released from the court of justice; these documents are dated and exhibit social realities – they are literary texts. To judge the designations of *chance*, it is necessary to further investigate the description for the game of dice.

In the Middle Ages, the game of dice required a minimum of three and a maximum of six players, and was sometimes played in teams; "two-player games tend to appear outside of real games, for example when two opponents rely on the fate of the dice to settle a dispute between them or to find a solution to a particular problem [...]. The vast majority of games is achieved using three dice. [...] The use of two dice is rarer. [...] The use of four is exceptional" (Mehl 1990, p. 96). The ability to score up to 24 points was not widespread and disagreement over scoring is one of the common reasons for brawls that took place as a consequence of the game. The most common goal of the game was to get the most points. If a roll was disputed, it was often when a die fell beside the track, or onto the ground, or any way other than straight, when it came to rest askew beside some obstacle, when it chipped or split: French people still say that the die is *cassé* (meaning "broken"), and less commonly the English equivalent is that the die is *split*. Cheating was not uncommon, and involved rolling the dice in such a way so as to get a specific amount. There were stakes; people used to place bets and there were professionals – François Villon was a professional dice player. The game of dice was forbidden on certain days, and, moreover, greatly frowned upon by the Church since it encouraged lying and brought about anger, the rationale being that lies and anger came down to harming oneself, *à se mépriser soi-même* which essentially translates as "to despise oneself". There were several varieties of dice games, which have different names, such as *la raffle, la griesche*[3] and *le hasard.*

The word *hasart* took on several meanings between the 13th century and the beginning of the 16th century. *Hazart* was used to point to the dice as an object, for a certain roll of the dice that resulted in a six, as well as a particular type of game.

Le Jeu de Saint Nicolas by Jean Bodel is a play dating from the beginning of the 13th century and stages a game of *hasart*. Three scoundrels play the "game of *hasart*" in a tavern with three dice. Only one player takes his turn and it is not known for sure whether the other two players form a group opposing the thrower, or whether one of the two is associated with him; the text is corrupted, but it is clear that Bodel's game of *hasart* aims for particular combinations. After Player 1's first throw was 13, Player 3, who is probably playing against Player 1, wants his team to take a *hasart*, which Player 1 was looking for; *hasart* therefore designates an amount, a number: 13 is not a part of this category of numbers.

> We looked for the meaning of the term *hasart* in the *Libro del ajedres de los dados y de tablas* of Alfonso X the Wise, King of Castile. This medieval game manual dates from 1283. *Hasart* refers to the totals of 3, 4, 5, 6, 15, 16, 17 and 18. In terms of probabilities, these totals are the most difficult to obtain. We understand that they ensure victory the first time. [...] The Spanish treaty specifies that if a *hasart* is not obtained on the first move, which is the case here with 13, the player continues to roll the dice. The total obtained at the first move then constitutes the "luck" of the opponents. [...] A *hasart* on the second move would lose the game [here to Player 1]. The latter then plays an 8, which constitutes his/her "chance". If he/she plays 13, he/she loses the game; if he/she plays 8 he/she wins it, which happens in the text around 1137 after a number of moves Bodel does not specify. [...] The game of chance therefore appears to be more elaborate than other dice games and, as a result, probably less practiced. We can therefore understand the rarity of its occurrences[4].

We will retain from this passage that *hasart* described different semiological material and intellectual layers of the dice game: (1) the object itself with the marked points; (2) the 6, the maximum number on the die and the number which allows one to win according to the rules; and finally (3) a game more complex than

---

3 *Griesche* means "Greek", as in "Greek fire", but instead here is a reference to the mythical inventor of the games, the Greek Palamedes.
4 From Mehl (1990, p. 94 and sq.). The "chances" for an 8 are less numerous than for 13.

others which includes at the beginning a statement of probabilities, a calculation of the possible combinations.

A poem from the 13th or 14th century and attributed to Jean Lefèvre, *La vieille, ou Les dernières amours d'Ovide*, shows the knowledge of probability in the game of dice (circa 1157–1175):

Tant de nombres puez trouver tu,
Tous ne sont pas d'une vertu,
Tous ne viennent pas egaument,
Car les moiens plus freqaumuent
Viennent que les grans ou meneurs.
De ce, sont les dez ordonneurs,
Selon fortune et meschéance,
Car une moyenne chéance
Bien souvent retarde et demeure,
Et petite ou grant vient en l'eure,
Si com quatorze plus tost qu'onze,
Dix avant sept, ce dont
Puet partout adviser.
En manieres cinquante et six
Sont en trois dez les points assis
Qui donnent diverses pointures
Selon diverses adventures.
Plus grant nombre n'y puet avoir.

It is evident that the author has reflected on the 16 number variations that a player can roll with three dice. To paraphrase the old poem:

This is the amount of numbers that you can draw, All do not have the same quality, All do not come out in the same quantity, Because the mean numbers appear more frequently Than the big ones or leaders. As a result, the dice are the orderers, According to good luck (*good luck*: a good roll of the dice) or bad luck (*bad luck*: a bad fall of the dice), Because an average result (*chéance*), Very often delays (the game), And a small or large result at the right time, Like 14 before 11, 10 before 7, which we can see everywhere. At any rate the number 56 Is with three dice the total of the possible draws Which give various results According to various possibilities, A larger number is impossible.

The author knows that with three dice, the total of possible combinations is 56, but on top of that, as any good *probabilist* knows, he recognizes that median combinations are more likely to fall than extreme combinations (Mehl 1990, p. 87). Let us note in passing that the author knew the *Liber Abachi* by Laurent de Pisa, alias Fibonacci, published in 1202 which introduced all the Indo-Arabic numerals into European practice, including the 0, hitherto excluded, alongside other graphical methods (Herrenschmidt 2007).

Hasart mainly ascribed certain *chances*, that is, certain outcomes, in fact certain numbers. The "game of chance" still existed in the 19th century and is documented in Larousse's *La Grande Encyclopédie*. It "knows many variations [...]. It requires two dice and opposes several players grouped together and represented by a single dice thrower and another who becomes the banker. Depending on a grid of combinations, the banker or the players win the stakes".

Let us examine another text on the dice games of the Middle Ages: the *Liber de moribus hominum et officiis nobilium ac popularium super ludo scacchorum*: "The Book of the Customs of Men and the Duties of Nobles or the Book of Chess", by Jacobus de Cessoles (Mehl 1995), an Italian Dominican monk who died in 1322. The passage that is of particular interest to us can be found in Book III (8, 368b): "Blessed Bernard (from Clairvaux), whilst out riding, met a player." The player said to him: "I would like, man of God, to wager my soul against the horse on which you are sitting, if you would like." Blessed Bernard replied: "If you pledge your soul to me, I will dismount. If you get the most points on the three dice, I promise to give you my horse for free." Happily, the man threw the dice and got 18 points. Confident, and believing in his victory, he took the bit from the horse, claiming it was his horse. Bernard said to him: "Wait a minute, son, the dots you see on these dice are more numerous." Indeed, due to the shock, the third dice had split in the middle: on the greater part there were six points, on the other one point. Thus, it happened that on the two other dice that Bernard had thrown, there were 12 points while on the third there were now seven, that is to say one more than in the player's case. Seeing the miracle, the latter offered his soul to Bernard and became a monk, living out the rest of his days in a befitting manner.

Bernard knew ahead of time that he would make more than 18 points – even at the cost of cheating. He knew there is no chance when God intervenes – faith saves one from chance.

In some cultures steeped in religion and religious thought, what we call chance may appear as "the action of a god (for instance Zeus) or whatever befits a situation". It is therefore fundamental to understand in these expressions of chance the evocation of a link, of a symbolic and necessary correspondence, which can be translated by the verb *to fit*.

The same could be said from the text of Jacobus de Cessoles above, where faith and revelation, supported by the cunning holiness of Bernard, are the negation of chance. Now it seems to me that this negation is here far more radical than the required correspondence of coincidental phenomena thanks to an invisible non-human power, evoked in the structure of words or in divine figures. In fact, with Bernard de Clairvaux, the unveiling of divine Providence not only eliminates chance from the story, but also denies it all the more strongly because it is done through a trick: normally, a "split" die annulled a game, as we have seen. There is no escaping the absence of chance. Everything is "linked" by providence, through it, a holistic symbolic world.

As we know, theologians saw in games of chance the work of the devil, which inspired the Roman guards at the foot of the cross to play dice with the garment of Christ. In the 12th century, the *Decretum Gratiani*, a compilation of divergent canonical texts, prohibited clerics from participating in games of chance; The Church tried to forbid them entirely, since dice could limit Providence. In the Greek language, *devil* is the antonym to symbol; the Greek *to symbolon* etymologically denotes "that which is put together", hence the symbol indicates an adequacy which links things together, for example a mathematical symbol and an operation. *To diabolon* signifies "that which is put across", hence the devil and connotes the disorder inflicted by this actor on the connections and correspondences between things.

At the same time as this hardening of the Church, the players, which included nobles and kings among their number, borrowed from Spanish the Arabic word which would later give the French word *hasard* ("chance"): in doing so, they departed from the vocabulary inherited from Latin or Greek. The notion of *chance* now slipped into a word unrelated to other familiar words in the vernacular and/or erudition. This is the first step in the process of dissociation.

The game of dice with its staging of chance allowed for certain calculations on the frequency of possible outcomes, which constitutes an additional aspect to the idea of dissociation which characterizes our thinking about chance. We can in fact assume that the perception and calculation of chances, as demonstrated in *Le Jeu de saint Nicolas*, the Spanish manual, and the demonstration by the author of *La Vieille ou Les dernières amours d'Ovide*, participate in the movement of certain preparations for the understanding of nature and the world in terms of numbers and calculation, to the detriment of Providence.

Science seeks causes and finds links between phenomena, which are not themselves "rewoven" by science, since what science states as truth is impermanent, always questioned, for the most part escaping the common language, the common symbolism characteristic of signs and reasons. It gave rise to the concept of determinism, which in many ways remains difficult to this day for many to accept.

It is through this perspective that the use of *stochasticity*, chance within the non-deterministic environment of biology, confirms the advent of a higher level of thinking when it comes to the notion of chance, a higher level in the same historical sense, that of dissociation: *stochasticity* tells us that the process of dissociation has already decoupled from chance. This higher level of dissociation puts into perspective the contemporary tremor affecting entire portions of our societies: for example, some people come back to the idea that the Earth is flat, because the dissociation that is commonplace within the universe of sciences, particularly that which engages with the living world, especially the human body, is too jarring for them and as such is unbearable.

So, in agreement with Plato, chance goes from the invisible to the invisible. But what do we place under this term, since we have removed from it "fitting into a situation, the action of God and the fight against Providence"?

What invisible remains to us, at the heart of our identities as scientists and citizens?

The one, I will answer, which guides both democratic societies and the sciences: the optimism that comes from free will.

3.9. References

Herrenschmidt, C. (2007). *Les Trois écritures. Langue, nombre, code*, 510. Gallimard, Paris.

Mehl, J.M. (1990). *Les jeux au royaume de France, du XIIIe au début du XVIe siècle*. Fayard, Paris.

Mehl, J.M. (1995). *Le livre du jeu d'échecs ou la société idéale au Moyen-Âge, XIIe s*. Stock, Paris.

4

The Collective Determinism
of Quantum Randomness

Chance exists. Everyone has experienced it in some form or another, for example, on those occasions of fortuitous coincidence that arise unexpectedly in everyday life. Chance is sometimes so helpful, and sometimes so maddening, that one is tempted to wonder whether it is completely random or whether it corresponds to a higher power.

> [...] but you did not foresee the death of M. de Turenne nor the cannon
> ball shot at random, which singled him out from ten or twelve others.
> I, who see the hand of Providence in everything, behold this canon
> loaded from all eternity. (Madame de Sévigné)

In physics, things are clear: we know the behavior of chance, which should not be confused with noise; it is not absolutely random, it follows distribution laws which indicate a collective determinism. In the world of the infinitely small, reality is probabilistic, and we will see that this result appears to necessitate a revision of our ways of understanding, meant here in the sense of our ability to predict.

4.1. True or false chance

A dictionary defines chance as the "cause of random facts, following one another fortuitously". This does not advance the discussion much. In classical physics, that is to say for events that arise due to gravity alone, there are no *a priori* assumptions for chance. For Newton, physics is strictly deterministic, that is to say that its manifestations are completely predictable and the law of cause and effect is the

Chapter written by François VANNUCCI.

cornerstone: an experiment must generate repeatable events in the same way so that the exact measurement will always deliver the same result. Determinism is therefore anti-chance, which has very satisfactory consequences for the mind: the reality of the world is potentially known out of time, since we can accurately predict the future through the application of dedicated equations; the falling stone is not free.

A contrario to determinism, we can therefore define chance as that which does not follow a strict law, that which will not be able to predict a future in the chain of phenomena.

It may turn out that a seemingly haphazard fact becomes understandable with advances in science. In fact, a large number of phenomena have found their rational explanation: in ancient times a solar eclipse foretold misfortune, until celestial mechanics were able to explain and predict it. The dice roll is the archetypal game of chance; however, if we carefully examine the conditions of the throw, with the aid of a fast piece of equipment (quite conceivable with current technologies), we are in fact able to predict the outcome. Therefore, is there or is there not the element of chance in the roll of a dice or the toss of a coin? I once read an article which made a point of showing that a buttered piece of toast has a higher probability of falling face down by the simple application of Newton's law, and assuming a drop of about a meter.

The notion of chance depends on our current knowledge, so it measures our ignorance. However, from the Big Bang to today, we roughly understand *grosso modo* the evolution of matter. The field of chance has narrowed and it has lost much of its power to inspire fear. If, in spite of everything, we speak of chance, it is due to the complexity of the world: the principle of cause and effect applies to each individual process chain, and yet we remain unable to keep up with the entire sequence of tangled facts, governed by too many degrees of correlated freedom.

> In the Mind there is no absolute, or free, will, but the Mind is determined to will this or that by a cause that is also determined by another, and this again by another, and so to infinity. (Baruch Spinoza)

To master the complexity of a situation, physicists invented the Monte-Carlo method, a name obviously reminiscent of the game of roulette. One simulates a phenomenon by injecting all the knowledge at our disposal, arbitrarily choosing from the relevant parameters, one after the other, through a "random" function, one then applies the known laws, runs the computer, and thus predicts the probable evolution of things. The method was born as a by-product of the Manhattan project, which built America's first nuclear bomb, and ever since then physicists have made unbridled use of it, as have financiers.

4.2. Chance sneaks into uncertainty

There does exist, however, a cause of chance that is intrinsically linked to the scientific method. It simply hides behind the uncertainties of measurement. Even gravity takes part in the "game" of predictions, since uncertainties exist both within the definition of the initial state and in the values adopted for the couplings which characterize the forces at play.

Indeed, wherever there is measurement there is error. For example, the mass of an electron is given by:

$$m(e) = 510.998918 \pm 0.000044 \text{ keV}/c^2$$

This is the result of numerous evaluations for which science has taken the weighted average. As new measurements are found, the error will decrease, but it will never be zero. Nothing is perfect in physics, apart from those specific values one decrees without any uncertainty. This is the case for the speed of light which has, once and for all, been stated as equal to $c = 299,792,458$ m/s. Not excluding the fact that this value is itself based on a measurement, in turn attracting its own uncertainty, the use of the metric unit: meter, which is supposed to be defined in an absolute manner. Given that the standard meter is a physical object, on display at the Pavillon de Breteuil, Sèvres, the precision of its definition is no longer satisfactory.

The decision was thus made to change this arbitrary choice, and associate c with the absolute. The uncertainty remains, but it took refuge in the definition of the meter. This sleight of hand is only the result of a convention, and in this case, we prefer to confer perfection on a universal size, independent of the human hand, rather than to an artifact made of a platinum and iridium alloy.

Measurement uncertainties therefore lead to insurmountable operational risks, which is not entirely without consequence. It is calculated that an error of just one meter in the position of our planet will later translate into an uncertainty of 10 million kilometers after 100 million years. Let us calm down, though, as that is just 10% of the distance between the Earth and the Sun!

> That is not all; then, you say, science itself will teach man (though to my mind it's a superfluous luxury) that he never has really had any caprice or will of his own, and that he himself is something of the nature of a piano-key or the stop of an organ, and that there are, besides, things called the laws of nature; so that everything he does is not done by his willing it, but is done of itself, by the laws of nature. Consequently we have only to discover these laws of nature, and man

will no longer have to answer for his actions and life will become exceedingly easy for him. All human actions will then, of course, be tabulated according to these laws, mathematically, like tables of logarithms up to 10^{-5}, and entered in an index. (*Notes From Underground* Fyodor Dostoevsky, 1864)

This sensitivity to initial conditions is reminiscent of the butterfly effect – often used in meteorology – the flapping wings here causing a hurricane six months later in the opposite hemisphere. Therefore, even a purely classical phenomenon, wherein only gravity is at work, is not absolutely knowable and never will be … technological progress may reduce uncertainty, but it will not eliminate it.

But all these considerations have remained within the framework of classical physics where gravitation is hegemonic. However, there are other forces at work in the Universe.

4.3. The world of the infinitely small

In the infinitely microscopic, the world in which particles are in motion, the dominant forces are electromagnetic, strong and weak. The gravitation which manages the phenomena of everyday life here is totally ridiculous. Otherwise, these forces respond to laws far more subtle than Newton's law, and quantum randomness will superimpose a level of reality that will only be known after its effective realization.

Chance, obviously, exists at the particle level: we cannot unambiguously predict the trajectory of an electron subjected to a force, but this chance is not arbitrary, it is constrained by probability distribution laws.

The quantum mechanics controlling these behaviors is a sometimes counter-intuitive branch of physics. It is based on two contradictory postulates:

– there is an evolution equation, the Schrödinger equation, equivalent to Newton's law $f = m\ \gamma$, which, although more complex in writing, remains deterministic;

– however, this evolution law is supplemented by a principle of reduction or collapse, at the time of measurement.

The Schrödinger equation does not predict the outcome of a measurement, but the probability of its occurrence.

A simple and striking example is given by radioactivity. We know of unstable natural elements, that is to say, those that live for a limited lifetime. This is the case for atomic nuclei that are formed from a combination of neutrons and protons, and where the number for one of those partners is excessive; we are of course talking about isotopes.

Let us take a sample of ^{14}C which is characterized by a lifespan of 5,730 years, suitable for dating ancient objects, in particular mummies. A nucleus taken at random evolves randomly: the first decays after an hour of observation, an absolutely identical second nucleus will change in nature after a year, and another after a thousand years. No one knows how to select the most resistant in advance. This process appears to be completely fortuitous, and yet, if we follow the evolution of the population as a whole, we see that the successive decays are distributed over a mathematically defined time curve. This is the so-called radioactive decay curve, an exponential one, an example of which is shown in Figure 4.1.

An individual nucleus behaves randomly, and a million nuclei generate a distribution that can be known in advance. In addition, the modes of decay may differ. Therefore, the meson π decays gives $\pi \rightarrow \mu + v_\mu$ in 99.99% of cases, but, in 10^{-4} of cases, it will decay in $\pi \rightarrow e + v_e$, without being able to anticipate the selected mode.

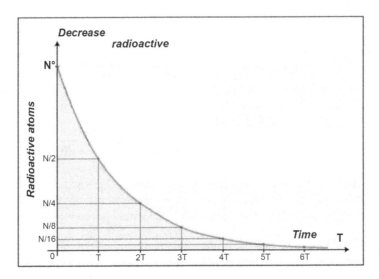

Figure 4.1. *Radioactive decay curve. For a color version of this figure, see www.iste.co.uk/gaudin/chance.zip*

4.4. A more figurative example

The interference patterns caused by the superposition of overlapping light have long been known. Young's double-slit experiment places two slits in the path of a light ray, which project onto a screen a shape with alternating luminous and dark fringes. Figure 4.2 shows the schematic for the apparatus and its result. The light is shared between the two slits. This is characteristic of wave behavior, and similar shapes have been found to occur on the surface of the water.

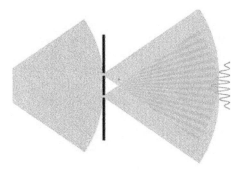

Figure 4.2. *Diagram of an interference experiment. For a color version of this figure, see www.iste.co.uk/gaudin/chance.zip*

This poses a problem on the nature of light, a debate which was topical as early as the 17th century. Newton's particle theory was opposed to Huygens' principle on waves. In the 19th century, Maxwell wrote his famous equations which unified electricity and magnetism, and associated the speed of light with two physical constants, characteristic of a vacuum. Mass appears to have been accounted for and the undulating character of the light proven definitively.

A significant drama of the 20th century took place in 1905, when Einstein interpreted the photoelectric effect as a collision between grains of light – which he called photons – with the electrons of a metallic surface. He went on to receive the Nobel Prize in 1921 for this major breakthrough which constitutes the first essential pillar of quantum mechanics, a discipline which he would later oppose fiercely.

Light is revealed to be of a dual nature, that is, both a particle and a wave. To be more precise, light is a wave when it propagates in space, but becomes a particle when it interacts, which occurs at the moment of production at the source, as well as at the moment of detection when it touches the screen.

The result is similar, and more troubling, with the electron, which sometimes passes itself off as a particle and or a wave. A beam of electrons is sent through a similar dual-slit structure, such as one that interferes with X-rays, i.e. extremely small. Will the electrons pass to the right? Or will they pass to the left? On the screen, a first electron "collapses" at a certain point. A second identical electron falls elsewhere, and so on. The point of impact is random, but a million electrons reconstruct an interference pattern, they distribute themselves in a predictable manner reproducing fringes, as shown in Figure 4.3. So, are these waves or particles? The electron is no longer local, when it propagates it is a wave, but when it interacts it is a particle. How can an electric charge or a mass be distributed in volume?

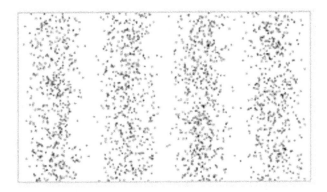

Figure 4.3. *Electron interference; each point indicates a particle impact*

From this example, we can see what a noise effect would be. There would be no structure and the screen would fill up continuously. Noise does not carry information, while an interference pattern provides information on the dimensions of the responsible network, which can be estimated accurately.

A single electron follows pure chance, a population draws predictable patterns of diffraction or interference, quantum randomness responds to a principle of "collective determinism". This property gives rise to various interpretations. The most drastic, proposed by Everett, rejects collapse: all states are objectively real and if we choose to only observe one end state, among all the possibilities, that is our own fault! A universe that has never been reduced to entangled states is brought into being, each possible state corresponding to an observer with their own individual perception. From one state is born many-worlds. What superabundance!

It would seem more economical to admit that there is a reality beyond our comprehension, but this is precisely what irritated Einstein and which caused the infamous verdict: "Quantum theory yields much, but it hardly brings us close to the Old One's secrets. I, in any case, am convinced He does not play dice with the universe".

In Everett's scenario, God does not play dice because he oversees the whole thing, but *we* ... do, through the act of observation alone.

4.5. Einstein's act of resistance

The legendary physicist believed only in absolute determinism; everything in the Universe had to follow rational laws, and his infamous choice of phrase reflected the suspicion toward indeterminism by quantum mechanics. He fought head to head against any and all advancements of the theory, which culminated in the EPR paradox.

In 1935, Einstein, Podolsky and Rosen published what they considered to be the decisive argument demonstrating the incompleteness of quantum mechanics, according to which two particles having a common origin cannot be considered as independent, referring to an entangled system. Any measurement on one reflects on the other regardless of distance. Einstein deduced that this *influence* between the two spreads faster than light, in violation of relativistic causation.

To remedy this problem, the physicist suggested the existence of additional hidden parameters that would describe these particles. Therefore, for the electron, we know the mass, the charge, the spin, but we must add a new quantity, yet to be discovered, which differentiates the electrons and will command one to go here and another to go there. There is, in fact, a discriminating quantity: the absolute time it takes for the consecutive electrons to arrive, however, the Schrödinger equation does not account for this parameter.

In 1964, John Bell suggested an experimental test; he wrote a series of inequalities giving the numerical relationships that correlated measurements must check while respecting causality. In 1982, Alain Aspect used a source that simultaneously emitted two entangled photons in opposing directions, and measured their polarizations at a distance of 6 m from the source, and 12 m between the two photons, thereby confirming the violation of Bell's inequalities, as predicted by quantum mechanics. Einstein's realistic local vision is invalidated, and quantum non-locality is demonstrated. The conclusion of the experiment is clear: there are no additional variables, one knows everything one can know about a process and yet one cannot know everything.

Heisenberg quantifies the uncertainties of quantum mechanics through his famous relations:

$$\Delta x \Delta mv \approx hc \text{ and } \Delta E \Delta t \approx hc$$

where x is the coordinate of position, v is speed, E is energy and t is time, while Δ quantizes uncertainty. Since Planck's constant h is not zero, it will not be possible to know exactly both the position and the state of motion for a given electron. It indicates more than uncertainty, this is indefiniteness.

A controversy ensued between Einstein and Heisenberg. For the believer in determinism, theory had to describe what Nature is. Heisenberg, representing the Copenhagen school, responded more modestly that: physics can only reason about what we know about Nature, showing how it behaves in a given situation. He adds that its uncertainty relations invalidate causality, since to verify this requires taking a measurement which disturbs the system in an uncontrolled manner. Causality is only valid "on average" and the role of the observer is crucial. The moral of the story: our knowledge will never be complete.

4.6. Schrödinger's cat to neutrino oscillations

So long as they are not measured, quantum systems can exist in superimposed states, each characterized by a probability of occurrence.

This sometimes leads to paradoxical situations. Hence, the parable popularized by Schrödinger which suggests that quantum mechanics allows for a cat in a box to be both alive and dead! A story so extraordinary it could have come out of the imagination of Edgar Allan Poe.

In his *La Scomparsa di Majorana* (The Dissappearance of Majorana), Sciascia exposes the problem:

> The disintegration of a radioactive atom can force an automatic counter to record it as a mechanical effect, which is made possible by an appropriate amplification. Ordinary laboratory devices are therefore sufficient to prepare a complex and conspicuous chain of phenomena that is *commanded* by the accidental disintegration of a single radioactive atom.

Where is the dilemma? As has been discussed above, decay is a random quantum phenomenon but the consequences are a matter of classical physics. Therefore, a microscopic event triggers an effect in the macroscopic world, one migrates from a

Quantum Regime to a Classical Regime, and this leads to the enigmatic fate reserved for "Schrödinger's cat". It is a devilish thought experiment which illustrates the apparently absurd path down which quantum mechanics can lead. A cat is locked in a sealed box that contains a radioactive atom that decays at the rate of 50% per hour. In the event of decay, a system releases a deadly gas that instantly kills the cat. What is the condition of the cat after an hour? Without opening the box, the answer is uncertain. The simple reason predicts a probability of ½ that the cat is alive and ½ that it is dead. However in quantum mechanics, this result is given by the square of a wave function, and interference is possible between the two states. This is not to say that the cat is either alive or dead and that one does not know it.

As soon as one opens the box, the wave function "collapses", and one finds oneself in front of a single event and the spell is sealed. However, before the examination, one cannot come to this decision. A troubling conclusion! According to the Copenhagen interpretation, the cat is both dead and alive. We can say that it lost its life, little by little, up until the experimenter opened the box. It sounds strange and yet this phenomenon is achieved with quantum objects. Putting a cat in a superposition of two mutually exclusive states is not feasible, but research with entangled photon pairs nonetheless supports the hypothesis.

Note that this situation could only be obtained in a world where time could be abolished. The cat always ends up dying; the dead cat and the living cat are two states which coexist when one places oneself "outside time". It is both dead and alive, like everyone else, if one integrates the function of time. The crux of the matter is how to be dead and alive at the same time.

Thinking further than the tragic state of the poor animal, sacrificed for science, let us examine the case of neutrinos which will give a more explicit picture.

There are three types of neutrinos: v_e, v_μ and v_τ. These are clearly distinguishable: the first gives an electron during an interaction; the second a *muon*; and the third a *tau*. But when they propagate they lose their individuality. After a certain amount of travel time, it is found that a neutrino, initially of the type v_μ, can interact like a v_e. This is called oscillation; a spontaneous change has occurred between one type and another type. What does it matter? Neutrinos interact so rarely anyway. But in the particle world, a v_e differs from a v_μ as much as an electron differs from a muon; or macroscopically, the difference between an apple and a pear. It is an entirely different object, and these oscillations prove the existence of masses affecting neutrinos, a problem that has been a nagging question for decades now.

The phenomenology is quite simple: we dissect the problem by introducing a set of three other states v_1, v_2 and v_3 to which we attach well-defined masses and which will therefore follow Schrödinger's equations of evolution.

The "physical" neutrinos will be linear combinations of mass neutrinos, the two sets being linked by a so-called mixing matrix of dimension 3×3.

Therefore:

$$\nu_e = \alpha \nu_1 + \beta \nu_2 + \gamma \nu_3$$

This is in fact approximate, since ν_1, ν_2 and ν_3 have different masses, the equation cannot conserve both the energy and the momentum of the neutrino. A rigorous calculation taking into account entanglement shows that the simplified result remains valid.

So what is the reality? In other words: who are you neutrino? ν_e or ν_1, ν_2 and ν_3?

The answer is dual: ν_e ... at the time of weak interactions – that is, at the production and interaction stages – but ν_1, ν_2 and ν_3 at the time of propagation which takes place at different speeds for all three components. A phase difference develops which explains the oscillation.

I know that a thousand billion neutrinos pass through my hand every second, they come from the Sun. What kind are they? The Sun only produces ν_e, but these oscillate as they travel and the flux received on the Earth is roughly democratically divided between the three types. If we follow a neutrino from the Sun, we can only say that, in terms of its passage on Earth, it has a chance of being ν_μ or ν_τ in nature. In order to be sure, it would be necessary to trap the neutrino, which is not easy. Since Schrödinger's cat is both dead and alive, the neutrino is ν_e, ν_μ and ν_τ with a certain probability of occurrence for each state of being. Nevertheless, in the case of my hand, I know quite precisely how many neutrinos of each type are passing through it. In a similar fashion, following experimentation with 1,000 cats, I will know that around 500 are likely to have lost their lives. The distribution of possibilities is always constrained by the laws of probability.

It should be noted that, when they leave the source, neutrino waves remain coherent over a certain distance, but far enough away, the ν_1, ν_2 and ν_3 no longer interfere. They are frozen and interact as ν_e, ν_μ or ν_τ depending on the percentages that remain fixed. Therefore, the Big Bang neutrinos, released one second after the initial explosion, exist in the form of ν_1, ν_2 and ν_3.

Here again, determinism becomes "collective" which provides a much weaker constraint than the strict determinism of classical physics.

Thus, chance occurs when we conduct a detailed analysis of phenomena, particle by particle, but there is no longer chance, strictly speaking, when analyzing a population. It is reminiscent of the law of large numbers which smoothes out the gaps. If I consider a crowd of visitors coming one Sunday to the Château de Fontainebleau, some will be there in memory of Napoleon, others due to the proximity of the forest, or visiting a relative living in the region, but the overall attendance will be pretty much predictable.

4.7. Chance versus the anthropic principle

Quantum mechanics is not limited to the infinitely small. After all, our Universe was born out of a quantum fluctuation and this raises the question of chance on a very large scale. Did we appear fortuitously? This key question, if ever answered, will have a meaningful impact on our future. Science provides us with leads; it lists the necessary conditions for the emergence of life, highlighting the fact that these conditions require a particularly favorable combination of circumstances. It is at this point that a new direction of thought invites itself onto center stage.

Physicists believe in the objectivity of Nature, yet some numerical coincidences seem too *ad hoc* to agree with our sense of truth, that is, the idea that what has happened is nothing particularly improbable. This observation prompts a rash reasoning which breaks the strict orthodoxy of cause and effect, referring here to the anthropic principle, according to which the world exists for our sake. From this standpoint, the why is no longer the *cause* but instead becomes the *objective*.

The Earth is very welcoming to us and life has been grafted onto it only by overcoming many obstacles. The Sun has the right dimension and radiates sufficient energy; its age as a star is likewise suitable. The size of the Earth is adequate and its distance from the star is satisfactory; its rocky nature has made it possible to retain the essential atmosphere for the greenhouse effect which preserves a clement temperature; the Moon allows for the tides that some consider as the source needed for the chemistry of life. We live on the Earth due to these very special conditions that exist. Unless we think that we are the product of some highly improbable coincidence, or some intelligent design, we must assume that our Universe contains a large number of planets and that the probability of life on one of these planets is no longer that unbelievable; a very large statistic is indeed a prerequisite.

The Universe contains 100 billion galaxies, each with 100 billion stars. The chances of finding a good planet are no longer negligible. We have already discovered more than 4,000 exoplanets and the hunt continues. Plausibility is preserved and the anthropic principle becomes null and void. Consequently, other planets here and there have probably fostered life.

However, the planetary argument does not respond to the adjustment of the physical constants that distinguish our Universe. Indeed, this one seems wonderfully tuned for our emergence. The parameters which dictate the behavior of matter have the precise values needed.

For example, the difference between the masses of neutron and proton is 3 MeV. Any smaller and the neutron would be quasi-stable and the stars would have exploded at an early age, the genesis of life would have been impossible. Any higher and the lifespan of the neutron becomes shorter and the Big Bang evolves in a drastically different way. It is the same for the mass ratio between protons and electrons, the value of the fine-structure constant or the speed for the expansion of the Universe; had it increased at a greater rate, the stars would not have formed; had it increased at a slower rate, matter would have condensed into dense black hole-type structures. A modest variation of one or the other of these parameters would have upset evolution and we would not be here to discuss the benefits.

There is no ultimate theory that derives the fundamental constants from the great principles. This is the Holy Grail that theorists have long been searching for, so far in vain. The values in play seem to be chosen at random, but nevertheless chance has done things very well. This observation gives rise to the idea of the anthropic principle which is based on the statistical difficulty of explaining these comfortable coincidences, the approach contests the sole cause of our arrival as chance and instead encourages contemplation on the possibility of Design. The secondary details of nuclear chemistry that determine essential conditions go against the plausibility of this line of reasoning, and instead invoke a new statistical argument.

In fact, a possible remedy that aims to be rational is proposed by the model of the Multiverse, which makes the anthropic principle teeter on the brink a little. Our Universe was born from a quantum fluctuation in a primordial vacuum, continued by a phase of very rapid expansion called inflation. But then, there is nothing to prevent a series of parallel fluctuations. In this scenario, an almost infinite sequence of Universes is created, because when the excited vacuum spawns a Universe the rest of the space continues to grow exponentially, and more bubbles form. A multitude of Big Bangs takes place, separated by rapidly stretched spaces and therefore without communication between themselves. However, inflation can occur in different ways in each of the Big Bangs. The constants take on various values that are impossible to predict. This seems to indicate that there is no larger principle at work. And thus, we exist in the Universe which offers the right set of parameters. Again the statistical argument avoids the idea of Design.

> Let me declare, only, that, as an individual, I myself feel impelled to
> the *fancy* – without daring to call it more – that there *does* exist
> a *limitless* succession of Universes, more or less similar to that of

which we have cognizance – to that of which *alone* we shall ever have cognizance – at the very least until the return of our own particular Universe into Unity. *If* such clusters of clusters exist, however – *and they do* – it is abundantly clear that, having had no part in our origin, they have no portion in our laws. They neither attract us, nor we them. Their material – their spirit is not ours – is not that which obtains in any part of our Universe. They could not impress our senses or our souls. Among them and us – considering all, for the moment, collectively – there are no influences in common. Each exists, apart and independently, *in the bosom of its proper and particular God.* (*Eureka*, Edgar Allan Poe, 1848)

The concept of the Multiverse, prophesied as early as the 19th century by Poe's imaginative mind, stems from the early musings on string theory, constructed from the standard model of particles. This theory aims to reconcile gravity with the other three interactions. All of this seems highly speculative and unfalsifiable. Still, in this scenario, our Universe is no longer characteristic of a global structure. Elsewhere, the constants and laws could be different, the number of dimensions different and at the origin of the laws, according to Poe, even God himself. The Multiverse theory invokes the idea of many Earths, and that we live in the Universe that is favorable to us. A bubble of life as we know it would remain statistically very rare, but the number of bubbles is almost endless. An advanced model estimates 10^{500} of such Universes. One wonders if this solution is more economical than imagining a unique and all-knowing God?

4.8. And luck in life?

What about the far less grandiose chance in everyday life? Let us return to the personal macrocosm to analyze the occurence of chance which manifests itself in the form of luck. A very special experience gave me the opportunity to attempt a statistical study which, although a little laborious, in the end proved significant.

Let us lay out the terms of the analysis: I walk a lot to comply with the wise precepts of a healthy life. I am observant, a natural trait for a reasercher, and subsequently I occasionally find money during my walks. If over a certain amount, it gives me a fleeting sense of satisfaction and I declare my day lucky; this is something that deserves mention in my private journal. For many years I have kept diaries which outline my ordinary exploits. From these, I extracted the distribution of lucky days as a simple statistical exercise in the hope that it proved to be something interesting.

Out of 11,700 days examined, there were 350 lucky days – according to my definition of luck – that emerged from the noise of my life! One per month on average is enough to move forward.

First I looked for correlations with the most obvious external data: the day of the week, the day of the month and the month of the year. Nothing to report. Not even for the 13th of every month. For my own information, I deduced that the horoscope is nonsense. Take note.

But the big surprise was hidden in the distribution of the intervals between two lucky instances, the results of which are shown in Figure 4.4. On the x-axis is the number of days elapsed between one instance and the next; on the y-axis the number of cases found. A law emerges: the curve follows a decreasing exponential. It looks like the behavior of a radioactive sample which decays over time. But watch out, here we are starting from zero after each instance. We cannot conclude that after a lucky day, fortune is increased, only the postulate of probabilities remains: the events remain independent and the average chance on a given day is always the same, $\frac{350}{11700}$.

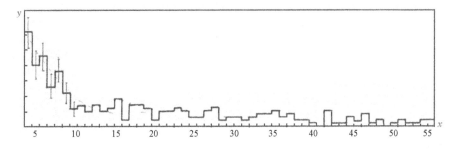

Figure 4.4. *Distribution of chance; number of days passed between two successive lucky days*

No matter what chance is quantified, it arises as a discrete event, and it follows a law of collective determinism! In fact, this distribution has been known since the great Pascal; it is called binomial, and it is difficult to imagine another. All the same, the result is disturbing ... the luck that we thought was completely fortuitous follows a mathematical law. The situation is in fact similar to that of electrons, which together draw an interference pattern, as if an electron knew in advance the distribution of its fellows that have already arrived.

It is the fault of not listening to the whole that we determine chance or irregularity for particular encounters. (Bossuet)

A mathematical law constrains the probability of chance, which only appears random if the events are rare, since a set follows a law. There emerges a sort of mathematical "transcendence", meaning here that which is beyond the intelligible.

It just goes to show that quantum mechanics has certain links with social life. The judgment by the "professional of chance", that is someone who lives as it were by the toss of a coin, becomes relevant:

> Fortune, which under the humbler name of luck seems but a word, is a very divinity when it guides the most important actions of a man's life. Always it has seemed to me that this divinity is not blind, as the mythologists affirm; she had brought me low only to exalt me, and I found myself in high places, only, as it seems, to be cast into the depths. Fortune has done her best to make me regard her as a reasoning, almighty power; she has made me feel that the strength of my will is as nothing before this mysterious power, which takes my will and moulds it, and makes it a mere instrument for the accomplishment of its decrees. (Giacomo Casanova, *Story of My Life*)

4.9. Chance and freedom

A corollary to the above stands out: in a strictly deterministic world, humankind would not be any freer than the falling stone, as warned Spinoza.

The stone is not free, but the electron has a choice, constrained, but real, and so do I. I found in Yukio Mishima's *Spring Snow* a reflection that fits here:

> To speak of chance is to deny the possibility of any law of cause and effect. Chance is ultimately the only irrational element that free will can accept.

> Without the concept of chance, the Western philosophy of free will could not have arisen. Chance is the chief refuge of will. And without it, the very idea of the game would be inconceivable, just as the West has no other way to rationalize the many setbacks and disappointments it must endure. It is my conviction that this concept of chance, of luck, constitutes the very substance of the God of the Europeans; they have a divinity there which draws its characteristics from this refuge so essential to free will, namely chance, the only kind of God who can inspire human free will.

So, does the individual have free will? The answer from quantum mechanics seems to be: yes, to a certain extent it can be seen as supervised freedom. The law of cause and effect only applies at the collective level but becomes invalidated – as Heisenberg warned – at the level of the infinitely tiny.

Nevertheless, we can take this reflection further. Freedom is what goes beyond necessity, so say the philosophers. Necessity makes one act: to feed oneself, to clothe oneself, to build a shelter, etc. But what is the point in knowing the age of the Universe or the mass of neutrinos? It does not matter to our daily lives, and indeed it seems that not many humans appear to be interested in such research. And yet a dark force pushes some of us to go beyond the frontier of that which is known. Why this need, given that we do not reap any immediate benefits from it? Research which only produces pure knowledge fundamentally proves our freedom. Einstein himself marvels at this tireless quest in which the Universe graciously allows itself to be examined. It is from the rationality of the world that he extracts his religious emotion.

> The most incomprehensible thing about the universe is that it is comprehensible. (Albert Einstein)

The world is understandable, okay, but only partially according to quantum mechanics. It forces us to conclude that our understanding of reality is limited. Leaving one with the feeling that ultimate knowledge is inaccessible to us. A no-nonsense argument easily reminds us of this: in this day and age, the physics of extreme matter comes up against the wall of gigantism. To move forward and to demonstrate the existence of the strings that fix the next step in the search for elementarity, a galaxy-sized accelerator would need to be built. Moreover, the limitation of our knowledge stems from another common sense argument: if our brains were able to understand all the secret workings of the Universe, as Einstein wished, that would require a very surprising coincidence between the scale of our intellect and the sum of the mysteries that irk us. The anthropic principle would return at full speed.

Our freedom is proven because our knowledge increases. Einstein wanted to know everything, which is very respectable. Our understanding of the world will continue to grow asymptotically, but sooner or later it will encounter insurmountable odds. As much as the reality of the world is disturbed by our observation, we cannot approach it in a neutral way. Note that our understanding is complete for the phenomena emerging on our scale. Our brain seems adapted to probe our close environment, but things become more blurry the further and more extreme the dimensions of the infinitely small and the infinitely large become, is that really surprising?

Spinoza only knew about gravitation; in this context everything is determined and we are not free. But then Heisenberg arrived on the scene and redimensioned our knowledge. He paved the way for deterministic physics, which resembles a meta-physical space which one can call, depending on preference, the Multiverse or God.

> In a word, if there is a god, all is well; and if chance rules, do not also be governed by it. (Marcus Aurelius, *Meditations*, IX.28)

5

Wave-Particle Chaos to the Stability of Living

5.1. Introduction

The human brain likes to make separations between objects, and within an object, between its different aspects, making it possible to easily distinguish and effectively communicate precise information. Human beings also like to create oppositions that allow for the transmission of binary information. This movement is balanced by an opposing secondary movement that likes to group common characteristics together. So, the ripe strawberry will join the Aigle[1] rubber boot in the red color class, even though it is a fruit (class of plants, subclass edible) and not a boot (a subclass of shoes).

The oppositions that can be queried here are mainly those of randomness/ determinism, and living/non-living (or physical), through the lens of consubstantial chance as seen through the classical presentation of Quantum Mechanics, an underlying deterministic chaos; the miracle that is life, even if it does seem so very fragile, the phenomenon of stability through quantum coherence. These questions also allow us to discuss the transition between an individual experience, a precise case (rough, affective) and an overarching statistical law (smooth, detached).

5.2. The chaos of the wave-particle

Aside from highly energetic particles and radioactive phenomena, the discoveries of which are relatively recent and not intuitive, the world through our

---

Chapter written by Stéphane DOUADY.
1 In France, the Aigle brand is well known for its red caoutchouc boots.

eyes seems to consist of a very stable substance, especially since chemists put forward explanations that there is no transubstantiation of matter, but rather a reorganization of atoms (and their electrons). If we look at the common stable particles: the proton, neutron, electron and photon, the latter seems to have attained a special status, one which, due to its nature, has long since been debated about: whether it is a wave or a particle. Louis de Broglie ended this controversy by explaining that the photon is in fact both a particle and a wave, and that either aspect can be considered according to the practicality of the experiment or the calculation in question.

By creating this wave-particle duality, which escapes the restriction of the preceding descriptions (in particular, their pseudo-incompatibility), Louis de Broglie demonstrated that all particles behave in this manner, even if the importance of their wave effect is often restricted and even forgotten or ignored. One of the consequences of this is the fact that the particle takes an extended position, over the width of the associated wave packet.

At the same time, at the beginning of the 20th century, the accumulation of results on Particle Physics (later referred to as Quantum Mechanics) produced an efficient formal model for the comparison of theoretical calculations with experience, by establishing the "wave function" calculations of particles (not to be confused with the previous *wave*). The wave function clearly represents a probability distribution for the particle to be in a particular place in a particular state. All that remains is to give it a meaning and an interpretation.

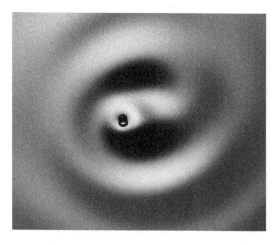

Figure 5.1. *Image of a bouncing drop inducing a parametric wave around it and showing the wave-particle object as a whole (source: Photo from Yves Couder's group, with permission)*

Oddly enough, the interpretation that has dominated (and still dominates) is that it is the particle itself that is random and indeterminate, as if it is in multiple possible states *at the same time*. This intriguing interpretation has sparked much discussion on the implications of such a fundamentally random reality. Most recently, it was this indeterminacy that launched a trend in quantum computing, that is the ability to compute all possibilities *simultaneously*.

A recent macroscopic quantum experiment, however, has revived another interpretation. The experiment involves letting a drop of oil fall into a bath of the same oil which is being oscillated vertically. As the drop of oil approaches the bath it crushes the air between the two, slowing it down. If one oscillates the bath at an interval period shorter than the typical coalescence time (a few tens of Hertz is sufficient), then the drop rebounds up into the air again and again, indefinitely. Upon hitting the surface, on its cushion of air, the drop not only spreads out, but also repels the bath oil, creating a small surface wave. Let us set the bath oscillation at an amplitude which approaches the Faraday instability threshold, where the vertical movement spontaneously generates surface waves (by amplifying noise). In this case, the wave created by the drop itself becomes a significant disturbance, remaining large by the proximity of the instability, before dissipating and disappearing as it is below the threshold. In doing so, a true wave-particle duality has been created (see Figure 5.1); a duality which is deeper than a common existence, since the very existence of the pair is conditioned by their common presence: without the surface oscillation, the drop would disappear through coalescence; without the regular percussion of the drop, the surface wave would likewise disappear through dissipation.

What is particularly fascinating is that this macroscopic wave-particle association can, under certain conditions, move spontaneously (the "walkers"): the bounce of the drop on the top of the wave that it creates becomes unstable, sliding onto the side of the wave. It then moves aside, but because it is on the slope of the wave, the next hit still ends up on the same side of the wave it has just created, and so on, propagating. In an enclosed space, or when passing through an opening, the movement becomes more complex, since it reacts with a wave that has been reflected by these obstacles (Figure 5.2c). In moving toward a wall with openings, these wave-particles also make it possible to reproduce classical interference figures, showing that they reproduce the character of both the corpuscular (the drop, which can be followed) and oscillatory (the wavy part that induces interference) (see Figure 5.2). The orbit of a small drop around a large one is also found, on average, to be quantified (by wavelength). In a potential trough (from a magnetic confinement of ferrofluid drops), we find the quantification for the orbits of electrons around an atom and the varied characteristics of higher energy orbits (see Figure 5.3).

In doing so, we realize that we can describe the motion of the wave-particle as *chaotic*, depending on the subtle interactions taking place between the wave, the particle and the external constraints. It is chaotic because it is the result of instability, for instance the walker, and the reminiscence of the wave. This creates a memory of the previous positions, which slowly disappears over time, and it is this memory which acts on the present moment. What this experience shows is that the greater the memory, the more the previous events occur; therefore, the more chaotic they are in practice, the more unpredictable they become. We become very sensitive to tiny modifications to the phase of the wave and of the particle, which is a parameter of the dynamics that has been overlooked because it was considered negligible. Here, memory plays a role as a creator of interferences, the ups and downs of which will orient the present in different directions.

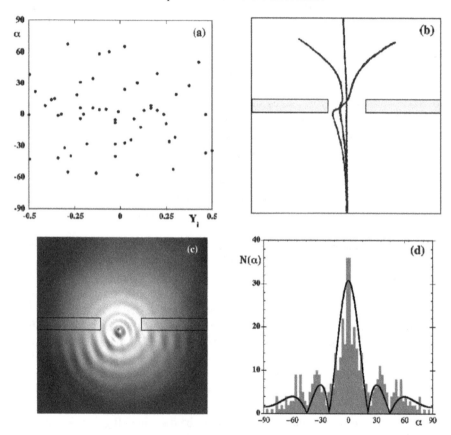

Figure 5.2. *(a) There is no relationship between the distance to the middle of the particle's slit and its exit angle. (b) For the same point of arrival (in the middle), we also observe very different exit angles. (c) Image showing the interference of a macroscopic wave particle through a slit. (d) The overall result is a classic, smooth, statistical interference (source: Yves Couder's Group with permission). For a color version of this figure, see www.iste.co.uk/gaudin/chance.zip*

Therefore, the extension of the wave and its memory provide richness to the particle's dynamics, which becomes non-local in neither time nor space.

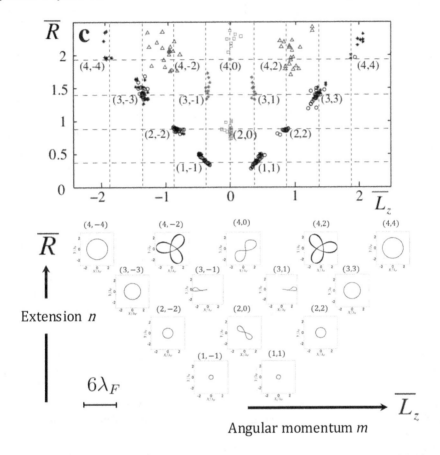

Figure 5.3. *Quantification of the trajectories for a macroscopic wave-particle in a harmonic potential. The first diagram represents the possible states observed experimentally, classified by their energy level (for mean radius R, extension) and their angular momentum L. The second shows the bits of quasi stable trajectories for each quantification island. All the trajectories are chaotic, unpredictably alternating bits of almost stable trajectories. These results are identical to those predicted by Quantum Mechanics, and used in Chemistry (s, p, d orbitals, etc.) (source: From M. Labousse, with permission, see Perrard 2014). For a color version of this figure, see www.iste.co.uk/gaudin/chance.zip*

To conclude, the apparently random movement of the drop follows not only a wave-particle duality, but also an underlying chaotic dynamic, due to both a memory of the system as well as the sensitivity of parameters forgotten as the phase.

Of course, this experiment does not perfectly correspond to a quantum system: energy is constantly supplied to a dissipative system, the wave obeys a surface wave close to a Faraday instability (and not the Schrödinger equation), etc. However, seeing this dynamic macroscopically, in particular its chaotic dynamic, gives a promising interpretation on the indeterminability of quantum particles that was previously not possible; symmetrically, it allows us to reconsider the quantum context: where does the energy of particles come from, their confinement, etc.

From this perspective, one returns to the existence of a well-defined particle in a very precise state, in which the wave function only describes an overall equation, a statistical equation valid for a set of possible outcomes. The need for this global probabilistic function would only arise from the chaotic nature of the dynamics of the particle itself: even if one knows it almost perfectly, it quickly becomes impossible to describe its state after a given period of time (extreme sensitivity to initial conditions, commonly known as the "butterfly effect").

It also resolves many paradoxes inherent to current interpretation. We start with the principal "measurement problem": if the particle is probabilistic itself, taking a measurement of it (for example, position or spin) provides a specific result. The act of measuring interacts strongly with the particle itself, "projecting" this measurement result onto it and "reducing its wave function". Measuring thus becomes an extremely disturbing action, which fortunately disappears after some time (with the system becoming indeterminate again).

This problem is best described by the Schrödinger's cat paradox, wherein a cat and some radioactive material are placed inside a box that has been fitted with a radioactivity detector, which on detection shatters a vial of deadly cat poison. Since the decay of radioactive particles is probabilistic (we only know the typical time it takes, but not the exact time when it will happen), the idea of measure amounts to saying that the cat is in an indeterminate "superposition of possible states", and only when the box is opened (upon observation) is the system forced to choose between the two states: dead or alive. This leads to many logical paradoxes. With the wave-particle interpretation, there is no longer a problem: the cat is either dead or alive, whether the box is opened or not. It is merely our inability to know the result in advance of opening the box that enables the experimenter (and theorist) to think of the scenario as a set of probabilities.

This interpretation is compatible with the absence (demonstrated, for example, by Alain Aspect) of a hidden variable, because we do not create a new variable that would be different for each particle. Well-known variables (similar to spin) are being considered here, whose wave nature (similar to phase) and its implications have not been realized. Bell's inequalities only come from their comparison between normal scalar variables, the probability of which one would know directly; and with wave variables, the probability of which is related to the square of the wave amplitude (intensity). If the "variables" are undulatory, their probabilities do not contradict Bell's inequalities (which were proven, for example, by Alain Aspect).

This also naturally solves the problem of entangled particles. They are twin particles, created at the same time, in the same place and in the same state, by a reaction. By sending them in different directions, after a while, we can observe their condition. In the current interpretation, looking at a state (such as spin) amounts to "projecting" the wave function, "reducing" it from a probabilistic set to a well-determined value. By making the same measurement, at a sufficiently close moment and over a sufficiently far distance, we find the same state for the twin particle. The common interpretation – since the particle is inherently a probabilistic wave function – is that the measure of one sends a message (how/why?) to the other particle to be in the same state. In other words, the wave function reduction is not only mysterious but also omnipotent since it is able to induce the same reduction on the twin particle (again, why?), acting faster than the speed of light. Experiments have also been carried out wherein a preliminary measurement is made on the first particle, which is then cross-checked with the other. In the current interpretation, this is because the act of measuring is itself an extraordinary perturbation (reducing the fundamentally probabilistic nature of the particle to a precise determined state). Finding the same state during simultaneous measuring proves that the first transformation has also been made on the twin particle, even when it has not been cross-checked.

In the interpretation of the chaos of the wave-particle, this is directly explained: the particles are in the same state because during their formation, one has in fact managed to put them in *exactly* the same state (which is impossible in Classical Mechanics, but possible in Computer Science). So, during their dynamics they will be in the same state at the same time. The mystery is then summed up by the fact that these particles are actually twins in an initial state, so close, that even subsequent chaotic evolution is unable to destroy their similarity. Likewise, the act of measurement again turns into a classic observation of the current particle state, without any modification. In this case, measuring it does not affect any changes and explains why one still finds the same state for the two particles later on. Once this is understood, we could look at the temporal dynamics of quantum systems, something which could already be looked into, given existing technological advancements, to reveal this chaotic behavior.

This example shows the difference between an individual description of a chaotic dynamic and the smooth description of the probabilistic set of possibilities. The fact of not knowing the origin of the indeterminate state of the particles naturally favored the global probabilistic description. However, there is a shift: when satisfied with a description of the world, the tendency is to substantiate it.

This description, which we have constructed on the very nature and explanation of the world, suddenly goes from a probabilist description of a set of possibilities to a single intrinsically probabilistic and indeterminate particle itself. What these macroscopic experiments show is that although they are not quantum particles in themselves, we are far from having truly explored what the wave-particle duality means, and in particular, the chaos involved. By directly visualizing these experiences, we can see the underlying chaotic dynamic, and understand its origin. In short, the "hidden parameter" of this indeterminacy is once again the parameter that is always forgotten, as Turing demonstrated with his universal machine: the parameter of history that allows a system to unfold its internal dynamics over time, and outputting the results as complex as needed.

5.3. The stability of living things

Living things always appear to us as being complex and mysterious, even almost more so than the quantum world in its probabilistic interpretation. Yet there is a parallel: what is shocking is the apparent unpredictability of the fate of each living being. It looks stable, almost unchanging ("Haven't I always been here always and forever more?"), but it can come to a screeching halt. "Accidents" and "sudden illnesses" are there to remind us of this fragility.

Figure 5.4. *A squirrel killed by a cat. Individually, it is an unforeseeable catastrophic event, but globally, these events are smoothed over by the continual evolution of statistical population curves (source: photo S. Douady). For a color version of this figure, see www.iste.co.uk/gaudin/chance.zip*

From a personal perspective, these events are dramatic. However, from a global statistical perspective everything smooths out. Although the death of a loved one is terrible, it can still fall within the normal statistical average. An abnormal case (an unusual crime) can be shocking, but is still statistically invisible. Important collective deaths, like a war, will almost always be less dramatic statistically, since they fall within the framework of natural disasters, which are beyond our control and to which we must therefore submit. They will be statistically visible as hollows in the age curves or as slight inflections to the overall population.

This illustrates how the living organism exhibits a stability that is surprising to us when it suddenly stops. Nevertheless, this stability is found globally: life *only* thrives on the Earth, despite natural disasters, like the one that followed the collision of an asteroid with the Earth, which initiated extensive lava flows, the combined result of which brought about the extinction of almost half of the known species present at that time.

Following this analogy, there appears to be a contradiction: in its complexity, living things (starting with a cell or even a bacterium) contain a large number of molecules. If it were to follow a global probability distribution, then that would be tantamount to saying that, globally, they should be in a more probable state, which is being mixed into the disorder and therefore becomes inert; like a set of gas molecules according to the second principle of thermodynamics. However, this is not the case and it remains to be understood why.

We are therefore faced with a whole: the entity, whose overall behavior impresses, without being able to relate to what is happening within. It is too tempting to project the notion of willpower onto this interior and fantasize about "little geniuses" who are acting consciously. Like the reaction to a "roly-poly" toy, which will start to rise and roll "spontaneously" after a period of rest, due to the movement of a little ball inside, which we do not observe the internal displacement of during the rest phases.

One of the first characteristics of living things is to "act in concert", to maintain oneself, via permanent movement. If we have the inner perception of being constant, we must never forget that we cannot go without breathing for longer than three minutes (on average), abstain from drinking for longer than three days, or fast for longer than three months. Without this our personal stability is brutally called into question. However, if we follow a reasonable diet (to which we could add a sufficient daily dose of sleep), we can maintain more or less the same stability, physically and mentally. It is maintaining these physical and psychic forms that allows this inner perception of constancy to be a part of our existence. In fact, what one perceives as constancy is the effort to keep alive, the effort to perceive, to perceive oneself. It is the constancy of a permanent movement. Obviously, this movement only exists, especially its internal perception, so long as the motion continues... but there is no reason why it should be endless. As the Buddhists say, [sic] *only that which is not born does not die.* In other words, what appears, because of a construction, however improbable it may be, has little reason to not disappear if and/or when this construction falls into disrepair. Our desire for its perpetuation is not enough, even more so if this construction seems highly improbable.

How does the living entity have this effective global coherence? This coherent movement allows for its maintenance, at least for a certain time. One possible answer, albeit partial, is precisely that contrary to the image of thermodynamic gas molecules, all the elements of the entity move "in concert", instead of returning to a state of disorder. How do we understand this?

If we go back to the image of microscopic physics, we can remember that its true foundation is not randomness, but is based on the fact that it is a world quantified. The variables can only take quantified values, limited to an integer number of possibilities. In a way, this is also observed in the living world. Put simply, an organism, like Schrödinger's cat, will indeed be either dead or alive. By erasing the difficulty to understand the passage between the two, which normally takes place quite quickly, but we now manage to extend in strange proportions with all medical techniques, the living is well quantified between two states. How is it possible?

One explanation is to just return to the movements in concert of all the parts of the entity, to return to the observation, and to say that it is precisely because they act in concert that we have a living entity. In a way, it is because all the parts are found to be in the same state, in the same collective motion, that the global entity (as a whole) is found in the same quantified state, which is frankly unique, that is to say: living. If we take the images of collective motion, we find this same phenomenon in the movements of sheep. A sheep essentially spends its time between two quantifiable states, either grazing or moving. It is difficult to imagine a sheep that is grazing (its head pulling up grass) moving forward at the same time: the mechanical cohesion of its body prevents this. If we assemble a flock of sheep, we generally find the same quantification: the herd will, generally, either be grazing or moving forward. It is easy to imagine that part of the flock is walking, and the other part of the flock grazing, but the desire to stay together makes this division unstable and the whole flock tilts toward a more homogeneous state in order to regain its cohesion (Figure 5.5). Recent experiments have even shown that these changeovers are more abrupt than those that can be deduced from the average changeover preference of a single sheep. The reason for this is that during the changeover, the individual sheep cannot be partly grazing and partly moving forward. This induces a more brutal global shift. If we imagine that the global state will also be forced to be quantified globally, in the same way that the mechanics of the sheep's body force it to be unable to graze and move forward at the same time, then we can imagine that this quantification can spread across scales, from the smallest (particle) to the global living entity.

Figure 5.5. *The collective motion experience, where a sheep trained to move toward a supply of food (red arrow) manages to attract the other 32 sheep along with it, for the sake of keeping the flock together (source: Toulet et al. 2015, with creative commons authorization). For a color version of this figure, see www.iste.co.uk/ gaudin/chance.zip*

In summary, the stability of the global living entity could be understood as a consequence of the phenomenon that is global quantification. This is what explains the appearance of the "individual".

5.4. Conclusion

In the case of the quantum particle, by forgetting its wave-particle duality, what is also forgotten is its own dynamic, its personal existence. Through the first necessarily difficult experiences, one is then only able to look at all the possible results, that is the overarching statistics. The major advantage of this is that it can be formalized and used in a practical way, without having to know what is happening at the individual level. The effectiveness of global smoothing has made us forget to seek out the underlying nature, and in cases even obliterate the very nature of the individual by superimposing the global statistical nature onto it. If we put this process off scale, as the experiment of Schrödinger's cat does, the absurdity of this process becomes glaringly apparent. The main interest of macroscopic wave-particle experiments is therefore to indicate not only what an underlying dynamic can actually be, but moreover, that this possible dynamic is ultimately much more "classic" and understandable than expected, thanks to chaos and its simple dynamics that are left to evolve over time. Nonetheless, nothing really prevents us, aside from our own conceptions, from looking at this individual happenstance directly.

In the case of living beings, the movement is reversed. By delving into the fascinating material details of the living object, one is struck by a complexity that one considers unattainable. We have tried to get out of this by going back to a formula of simple coded information that would (how?) allow for the control of all the details. Therefore, deferring the problem of knowledge to a molecular structure that knows everything. One might instead insist on the coherence which makes its unity, and which could come quite simply from an overall cohesion. Staying together, even mechanically, implies a certain harmonization, and ultimately a certain quantification of the possible states. In this case, the chaos of all the particular possible movements is erased, not by a global statistic, but by a coherence of all the parts to form an individual entity in a quantified state. Chance, the underlying chaos, remains visible only over very long periods of evolution, where it is found in the swarming of life forms and species.

In both cases, there is the difficulty of managing chance. The first comes down to imagining it as the very substance of the individual particle, smoothing it out and denying its possibly chaotic origin. The second amounts to making this chance disappear, not in a statistical way but in a concrete way, in the specific case of each living individual through quantum coherence. Integrating this into the description would perhaps allow for a more realistic, or at least serene vision of ourselves and of the world, by better understanding and accepting these moments of decoherence when everything, including us, can disappear.

5.5. Acknowledgments

To the members of the Yves Couder team, in particular Emmanuel Fort and Matthieu Labousse for sharing data, to the members of the Louis de Broglie Foundation for their stimulating meetings, and to Bernard Hennion for his friendship and inspired proofreading.

5.6. References

Perrard, S., Labousse M., Miskin, M., Fort, E., Couder, Y. (2014). *Nature Communication*, 5(3219), 201.

Toulet, S., Gautrais, J., Bon, R., Peruani, F. (2015). *Plos One*, 10(10), e0140188.

6

Chance in Cosmology: Random and Turbulent Creation of Multiple Cosmos

With no logical pretence, I dedicate this cosmic narration to Marie-Christine Maurel, whose songs of free scientific speech are vivid, and whom, I have built hotels for foreign nucleic acids[1] with.

What follows is less about the essence of theoretical cosmology and more about a large peripheral zone that is its telling outer casing.

6.1. Is quantum cosmology oxymoronic?

Something that is vibrant, alive, will come forth from the Cosmos, the order and finery of the Universe: a star, a cell, a thought, a "Bubble Universe"... at every second, this expression finds affirmation. The science of *Urania* is in its golden age as the grand discourse of the world unfolds, like an ancient summer, its quantum verses riding on scintillating tides. Going back to the origin, the inner school has rejoined the outer school in the vicinity of zero-time, zero-space, that damned initial *singularity*[2] (an euphemism for absurdity), communicating its confusion and perverse uncertainty in the process. The particle accelerator has become an astronomical tool for better or for worse.

Knowledge of space, time and matter, which not so long ago was objective and precise, has become uncertain, contingent to and dependent on the observer. I am

Chapter written by Michel CASSÉ.

1 Odile Jacob, *Xénobiologie*.
2 Reverse the arrow of time and the universe contracts to point at which it reabsorbs itself.

not talking about the multiplication of the number of space dimensions that threatens near zero-time[3], nor the transmutation of real-time into imaginary-time[4].

Loquacious, imaginative, disheveled and megalomaniac; at times modern cosmogony exaggerates; it exaggerates by pretending to rewrite the genesis in physical and materialist terms. All the same, we should not be moved by this, the scope of its knowledge and the ontological and existential questions it addresses. Let us leave the great logos, the interminable discourse of the origins unshackled, in the safe knowledge that pure epistemological prudence will be cast by Jean-Charles Pomerol, Jean-Paul Delahaye, Gilles Dowek and Yvar Ekeland onto the marble of cosmology, dissolving any stains.

Accordingly, the physical theory on the birth and infancy of a washed and sanitized universe shall not abuse the language of, nor intrude upon, the territory of the Judeo-Christian-Islamic God; we do not seek any quarrel with Him[5].

The first task of the astral physicist is to purify language as much as possible. It is necessary to be clear about the game from the outset, and to distinguish between theology and cosmology, physics and metaphysics, even though in the vicinity of zero-time it is difficult to detach observation from theory and cosmology from cosmogony. This distinction is, as we shall see, nevertheless crucial since *Inflation* (the exponential expansion) is not *Genesis* (the exit from zero-time). The first belongs to the category of evolution, and the second to that of creation. The first precedes the second in the same manner that one precedes two. One is original, and by virtue the other cannot be. One is sharp intelligence, the other broad intelligence.

Within a bellows, between the nozzle and the lung, there is the intake and there is air; *the* origin (of the universe, stars, life and consciousness) is always more obscure (pointed) than the evolution (broad). To this end, mathematicians and theologians must forgive our indecisiveness on creation as it is irretrievable. Around the inauguration of a cosmological discourse, the *narrator* is in a coma; then, at the base of the unconscious ideas become confused and the philosophical *aporias* block the march of reason toward its original point: zero-time is a moment in time that does not yet exist. In actual fact, zero is too accurate and precise a nomination to be quantum. The mouth of the universe has a cloud of uncertainty 10^{-33} cm in size, into which light would take 10^{-43} seconds to pass through, which makes no sense since in this minuscule time frame (as calculated by Max Planck), time is not precisely defined, and it may as well be called eternity. The age of the universe, estimated at

3 The consequences of Superstring Theory.

4 Steven Hawking, *A Brief History of Time*.

5 There is genesis and there is murder, the assassination of antimatter.

13.8 billion years[6], is thus knowable to an infinitely close temporal measure. There is no greater uncertainty in this world.

More often than not, Planck's frankly absurd and incomprehensible first 10^{-43} seconds (Planck's time) are quietly left to one side and the 13.8 billion years that follow are preserved. Here, we will do the opposite. We will store away the *gangue* of history and hold on to the gold that is the creation.

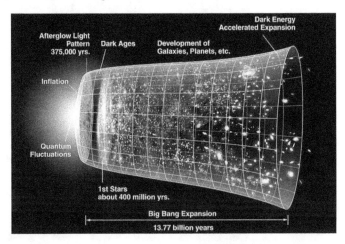

Figure 6.1. *Evolution of the universe since the emergence of classical time (the time of discourse). Quantum is synonymous with fluctuation and the first to enter the world is that of quantum fluctuations that will serve as the seeds of galaxies (credit: NASA). For a color version of this figure, see www.iste.co.uk/gaudin/chance.zip*

Time was born with the sky, time was born with the world, contemporary cosmology does not refer to anything other than Plato and Saint Augustine, but adds a *codicil*: What is born to knowledge and to narrative is "classical time", human discourse: the epic of Odysseus; the Bible; and of the Upanishad, where it has been associated with the "space of small-tiny birds".

We are often told that before the Big Bang there was absolutely nothing, and that this was the absolute origin of everything, including space-time. Before the beginning (of the discourse, one should hasten to add) there was neither time nor space. Can one measure the absurdity of this allegation? The "Before" is a temporal notion, as in the phrase "before the flood". We therefore reason in a circle. A slight straightening of space-time is enough to put it back on an even keel.

---

6 Determined by the Planck club, this time round by the satellite mission. See the *European Space Agency* Website. Planck is everywhere, in the sky and in the *oven*. It is at the origin and conclusion of cosmological discourse, at the beginning with its constant and at the end with its mass.

Figure 6.2. *Speculative cosmology. There is no better illustration of Quantum Cosmology than this woodcut by Flammarion, unknown author, Paris (1888). Colorization: Heikenwaelder Hugo, Vienna (1998). The character puts his head behind the sky. Nietzsche would call him the "hallucinated of the Back-World", akin to Plato[7]. For a color version of this figure, see www.iste.co.uk/gaudin/chance.zip*

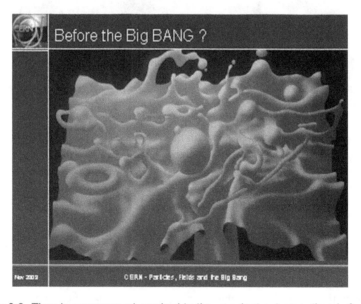

Figure 6.3. *Time is compressed, pushed to the very last extreme, time is foaming. For a color version of this figure, see www.iste.co.uk/gaudin/chance.zip*

---

7 https://en.wikipedia.org/wiki/Flammarion_engraving

To straighten out the semantics, we must hammer home the fact that the poorly named[8] Big Bang is not the origin of the world, but rather that of the Word, and the quantum uncertainty attached to it is not in the world, but in us.

Enough with metaphysics, let us instead write a few words and etch a few temporal notches on the equations and charts of the cosmos. In short, let us sketch out a chronology.

Figure 6.4. *The visualized arrow hits the target before leaving the bow. The force that bent the Bow of Odysseus is indeed intact and well in the hands of cosmology. For a color version of this figure, see www.iste.co.uk/gaudin/chance.zip*

Before the pistol shot, marking the launch of the universe and the official start of the cosmos marathon, there was no *classical* space or time, nor was there matter. Time was foam and vegetation, matter could not express itself. At least that is what they say. Today we are in a happy time, when matter speaks, where the mind can decipher the birth registration of the universe and put the records of the cosmos in order.

Cosmology is in its golden age thanks to the astronomical conjunction of telescopes, particle accelerators, computers, natural human intelligence and artificial intelligence.

---

8 A misguided onomatopoeia insofar as nothing can detonate a pre-existence environment, nonetheless, an excellent theory which has lent it name to a remarkable television series (*The BigBang Theory*), which helps to make fundamental research pertinent.

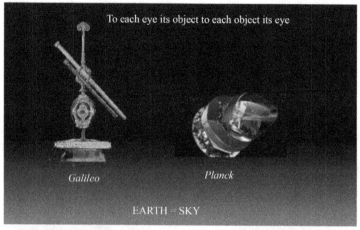

Figure 6.5. *(a) To each eye its object to each object its eye. (b) SNO: underground neutrino detector (Canada). For a color version of this figure, see www.iste.co.uk/gaudin/chance.zip*

The revolution is in the sky. As it turns out, Urania, the Greek muse of astronomy has been grasping at straws and mankind has learned more about the sky in the last 30 years than in the last 3,000 years. But each clarity engenders its own mystery: an invisible energy (said to be deceptively black) has taken the reins and is whipping the universe which is accelerating its pace. The elements of the universe are now three: Matter, Light and Dark Energy[9]. Thus the genesis is written, with time being counted from the beginning of the expansion of space, the mirage of the origin.

$$t = 10^{-43}s$$

Figure 6.6. *Virgo: open-air kilometer gravitational wave detector (Italy). For a color version of this figure, see www.iste.co.uk/gaudin/chance.zip*

With the birth of classical space-time, the radius of what will become the observable universe is tiny: 10^{-33} cm, the temperature is 10^{32} K and the density is 10^{94} g/cm3.

$$t = 10^{-35} s$$

The universe in a state of "false vacuum", brimming with energy, expands extraordinarily in a very short time (on the order of 10^{-32} seconds). This unimaginably brutal expansion is called the Primordial Inflation[10]. Today, it takes on the figure of Aladdin's lamp, from which universes emerge, as we will see shortly.

9 Cosmological constant, quantum vacuum, quintessence?

10 Criticized by one of its creators, Paul Steinhardt, who offers an alternative.

This is what cutting edge cosmological research dares to claim. The dimensions of the universe are multiplied by 10^{50} or more.

At the end of the inflation, powered by a mysterious entity called *Inflaton* (assuredly, a scalar field[11]), the temperature of the universe, which had fallen due to the expansion of space, increases again to 10^{27} K. We would then declare a Big Bang. After that classical cosmology follows and takes over. It is under the control of Particle Physics proven in the laboratory at CERN.

$t = 10^{-11}$ s

Radioactivity[12] and electricity separate. The universe is roughly the diameter of the Earth's orbit, its temperature is 10^{15} K and its density is 10^{33}.

$t = 1$ s

The temperature has dropped to 10 billion degrees and the density (mostly light) to 1 million g/cm^3. Neutrinos separate from matter to talk to François Vannucci, before living their free neutrino life.

$t = 380,000$ years

Figure 6.7. *Diffused cosmological background captured by the European satellite Planck. It is stirring to see the first forms emerge from out of an undifferentiated substrate. For a color version of this figure, see www.iste.co.uk/gaudin/chance.zip*

Nuclei and free electrons combine to form atoms without being sliced by photons. The attachment of electrons leaves the path free for photons; the universe becomes transparent to its own light. Light is released. This is the birth of astronomy.

---

11 Neither spin, nor vector, nor tensor (of spin ½ 1 or 2) but zero.

12 β.

The rest of universal history belongs to astrophysics: cosmic clouds give birth to star lines that open like flowers and swarm in a space made up of legions of winged atoms.

Figure 6.8. *Infrared Spider Nebula. 10,000 light years away in the Cocher Constellation, and everywhere where there are large interstellar clouds, stars are born (credit: JPL-Caltech, Spitzer Space Telescope, 2MASS; APOD). For a color version of this figure, see www.iste.co.uk/gaudin/chance.zip*

Let us take a step back for now. Cosmology, the science of the universe as a whole, therefore one of the greatest of thinkable ideas (*logos*) whose reigning force is gravitation, is based on General Relativity; elegant, abstract, confident and hellish deterministim, while the complex, uncertain and disturbed microcosm is under the thumb of Quantum Mechanics uncertainty. The two theoretical caryatids face each other like a pair of faience dogs at an impasse, arguing all the time, over time.

General Relativity considers time as an accessory, even illusory[13], while Quantum Mechanics considers it as absolute, sacrosanct and universal[14], which makes it regress to Newton's time[15].

---

13 It is flirtatious to say that it does not even exist.

14 It accredits certain instantaneous interactions at a distance, all the while claiming that no benefit can be derived from them (see the EPR paradox).

15 Quantum field theory, mother of the Standard Model of Particle Physics, reconciles QM and RR, but GR remains reluctant to quantify.

Yet the peaceful coexistence of the two queen theories, seemingly black and white, each imperial in its domain and never compromised, is absolutely necessary if we want to unravel the secret of the origins, because at the beginning of cosmic history, when the universe was very small, no bigger than an atom, the two worldviews are called to the bar; the bar of the lofty caravel of cosmology that threatens to get lost in the fog of time.

Classical cosmology suffers from an incurable evil brought on by General Relativity, that of an infinite birth. Give us an aberrant, magical origin and an admissible evolution, with computable Mathematical Physics and everyone will be happy.

If we want to avoid this resignation of the mind, this surrender in the open country of reason, we must by all means avoid or exterminate the "primordial singularity", a euphemism of madness or the shipwreck of the ambient theory, which compresses the entire universe to a point without thickness and arouses the horror of infinite density.

To do so, if we wish to indulge in Quantum Cosmology, it is imperative to generalize Quantum so as to make it compatible with gravity.

There is a certain style in Quantum, chapels and obedience, marked by divergences in interpretations, each more baroque than the next: the Copenhagen[16] fashion, the Everett[17] trend, hidden variables[18], Relational[19] or QBiste style[20] and so on.

The Quantum that we read about in books is not compatible with cosmology, and for good reasons:

1) it gives an immeasurable role to the observer and to the act of measurement, yet there is no one in much of the history of the universe to measure anything;

2) it does not allow for post diction since the act of measurement engenders the fundamental, irrepressible deadly peril for the bird which wants to find its nest, thus the irreversible and absolutely random reduction of the wave packet (collapse).

---

16 With incomprehensible and random reduction of the wave packet.
17 With disturbing multiplicity of worlds.
18 Hardly drinkable Bohm-de Broglie recipe.
19 Dear to Carlo Rovelli.
20 Taught by Chris Fuchs.

I have given the reason for the cosmic bankruptcy of the microcosmic vision of the Copenhagen Interpretation, which is otherwise wonderfully effective in the laboratory; I leave the technology to the wizards of Quantum Cosmology[21]. I focus on the scientific rewriting of creation.

There is a physical chain of the Vacuum-Light-Matter-Vacuum genesis. Empty or quintessence, the heart hesitates[22]. In this genesis, the light does not come first. It is in relation to the vacuum what the daughter is in relation to the father.

And for starters, the beginning never ends with being built.

An important difference between them stems from the fact that the smaller the space-time intervals under consideration, the greater the uncertainty about the energy involved in the phenomena. In fact, time and energy, so-called complementary or conjugated variables, in quantum *cenacles*, are partly linked, and as such they are subject to a relationship with uncertainty of the variety Heisenberg[23] was so fond of.

Written in full it is $dE.dt \geq h/4\pi$, where h is Planck's diminutive constant, small in size but large in mind and dE and dt express uncertainties about energy and time.

The Time-Energy Uncertainty Principle allows for all fantasies, on the condition that they respect the relation of uncertainty and of the electric neutrality of the vacuum, for example, the creation of photons (of zero charge) or particle-antiparticle pairs, more or less joined, then separated, then reunited again. These ghostly entities are said to be "virtual". Very useful, they are as real as you and I, much more fleeting, but no less estimable.

Virtual, or rather vacuous entities, since they emanate from a vacuum, are in common use in Quantum Field Theory, the possessive mistress of the Standard Model of Particle Physics which should be counted as a wonder of the world, a treasure of humanity. If the real and durable particles – quarks, electrons and neutrinos – are the building blocks of matter, the vacuous or dream particles (theoretical, but not hypothetical with the exception of the graviton) that carry forces and transmit them from one real particle to another are the bedrock. The

---

21 S. Hawking, J. Hartle, A. Guth, A. Linde.
22 In any case, not Einstein's cosmological constant since it has an extremely short lifespan. Let us say scalar field, everyone will be happy. Rendez-vous at Sean-Carroll's (blog) for tasting.
23 It would take too long to explain, so go ahead and open your quantum mechanics books. That of Jean-Marc Levy-Leblond and Françoise Balibar are certainly among the best.

vacuous/virtual particles put the real particles, the material support structure of everything, in relation.

6.2. Between two realities – at the entrance and exit – is virtuality

Figure 6.9, which bears the Feynman signature, describes the kiss of death between an electron and a positron, doubly antagonistic and deadly for the former (as it is charged more). This microcosmic conflagration gives birth to a so-called virtual photon that decays into a quark-antiquark pair, with the antiquark spitting out a gluon[24]. The jagged line expresses the hidden contact process peculiar to quantum physics, which happens so fast that it is not seen.

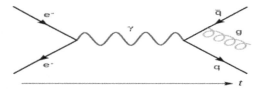

Figure 6.9. *A scene showing microcosmic love. e– and e+ represent, respectively, an electron and a positron (antielectron), γ represents a photon, q and q– represent a quark and an antiquark and g represents a gluon. For a color version of this figure, see www.iste.co.uk/gaudin/chance.zip*

6.3. Who will sing the metamorphoses of this high vacuum?

The four forces[25] are in the vacuum. There are as many types of vacuum as there are forces. The Absolute Vacuum, the sum of all Partial Vacuums, thus appears as a bank and a public relations agency.

The vacuum, hereafter nourishing fare, is too rich in terms of energy. Its energy is through the roof. It is so crowded that one becomes surprised to see it through. It is the subject of physics' greatest scandal, the so-called cosmological constant: its (energy) density, delicately calculated by our excellent friends in particle physics, is 10^{60} times greater than that measured by astronomers. There is something rotten in the realm of physics which reflects the hiatus between General Relativity and Quantum Physics.

---

24 Its birth did not go down well.
25 Electromagnetic, strong, weak and gravitational.

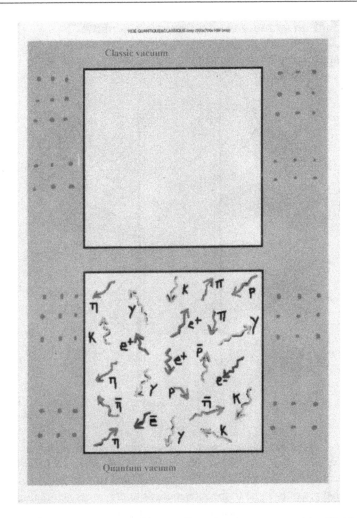

Figure 6.10. *Flowery vacuum. Above: the classic vacuum identifiable by its nothingness. Below: the quantum vacuum, which looks more like a Persian market than the Sahara desert. It is full of virtual particles that take advantage of energy uncertainty in order to exist for a brief period of time, the shorter these are, the more massive they are. For a color version of this figure, see www.iste.co.uk/gaudin/chance.zip*

6.4. Loop lament

Time, as stated, does not matter: for Quantum, time is absolute and existing (Newtonian), for the Relativist it is of course relative, even illusory, to be frank, it is Einsteinian. How do you want the two cosmos, micro and macro, to become one?

6.5. The quantum vacuum exists, Casimir has met it

The quantum vacuum is not something that exists only in the mind. The Casimir Effect has beared witness to this[26].

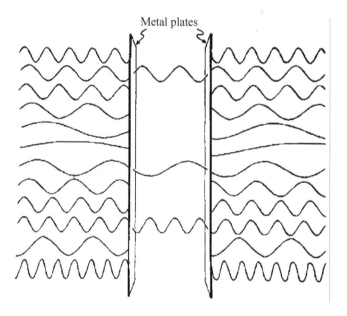

Figure 6.11. *Casimir effect*

Let two electrically neutral metal plates be brought under a demanding laboratory vacuum. The electromagnetic waves of the vacuum (quantum in this one), or if one prefers, the virtual photons, are selected so that the distance between two successive peaks is equal to the interstice, or to one of its sub-multiples. This is the effect of quantification. It has more waves or photons outside than inside. The external pressure causes the plates to come together. This effect is observed and scrupulously studied in the laboratory[27].

6.6. The generosity of the quantum vacuum

The vacuum is not a thing, it is a stable or moving state, like destiny. The "*true vacuum*" is the minimum state of energy of a system, and there are as many vacuums as there are systems, an electromagnetic vacuum, a weak vacuum, a strong

---

26 As well as the Lamb shift and the vacuum polarization, which we will not discuss here.
27 In particular, that of the École Normale Supérieure.

vacuum, etc. And besides the true vacuum, there are the false ones, angry, excited, unstable, but always connecting.

Figure 6.12. *Strong vacuum. We must see the gluons, half-wave half-particle virtual entities which circulate between the quarks! (Credit: Franck Wilczek). For a color version of this figure, see www.iste.co.uk/gaudin/chance.zip*

The quantum fluctuations of the quantum vacuum, or more exactly, of the quantum vacuums have become the handymen of physics, the good faeries, and the list of their benefits grows as the good works of Inflation to whom they are the godparents to do. And the blackhole, smiling, caressed by Hawking, begins to shine.

The initial Great Vacuum which marks the beginning of the history of matter is still called the *Inflaton*, a powerful quantum motor driving Primordial Inflation, while the current Small Vacuum, or *dark energy*, of the same nature as the Great Vacuum, serves as an accelerator for cosmic expansion[28], as discovered through distant supernovae[29]. The Great and Small Vacuums are "space spacers", or if you prefer, explosives. The Great Vacuum (Inflaton) was emptied of its energy and became light, and the light split into matter and antimatter. Out of respect for the vacuum, the most always goes with the least, and only zero is lonely. By this I mean that the photons (virtual, with zero electric charge) emerge from the vacuum

---

28 Attractive gravitation, the attraction of matter for matter, acts as a brake.
29 2011 Nobel Prize.

individually to return there immediately, while the electrons with less charge are always accompanied by a positron at the exit of the vacuum and again at the re-entry.

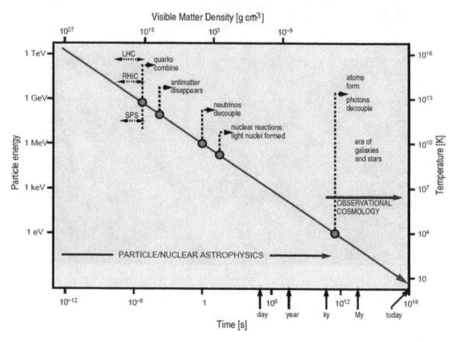

Figure 6.13. *The history of the universe. There is a physical and thermal chain of genesis: Great Vacuum – Light – Matter – Small Vacuum. The universe is an oven that was extinguished 13.8 billion years ago. Phase transitions (qualitative jumps) follow one another as the cooling induced by the expansion of space takes effect. Events are no longer monitored above a temperature of approximately TeV, i.e. within 10–11 seconds of zero-time convention (LHC, RHIC and SPS are names of particles). For a color version of this figure, see www.iste.co.uk/gaudin/chance.zip*

Last century, around the 1920s, physics underwent two revolutions, that of Quantum and that of the Cosmos: the discovery of the quantification of action (mass multiplied by speed multiplied by distance) on the side of the Earth, and of the expansion of the universe, on the side of the Sky. Mankind has not yet recovered.

Space expands between clusters of galaxies at a rate of 70 km per second per megaparsec[30].

---

30 1 parsec = 3.26 light years.

The universe has been an oven that has been extinct for 13.8 billion years and as the cooling induced by the expansion of space expands, the "symmetries" are shattered one by one. The original superforce is cut into four, like hair. It is cut in half and then halved again. The CERN LHC allows us to explore history and describe with confidence the events that unfolded beyond the 10^{-11} seconds, counting from the start of space expansion[31], but not before. Between 10^{-11} seconds and the mythical time zero, there is a chasm, especially since we can no longer comment on anything below 10^{-43} seconds (Planck time).

A closer look at Planck's time: the notions of time, distance (space), mass and cooking temperature are interconvertible through the following relationships, and hence express the solidarity of the physical genre.

$$x = ct$$
$$E = mc^2$$
$$E = -Gm/r$$
$$E = h\nu = hc/\lambda$$
$$E = kT$$

with, presentation of the characters: E = energy, m = mass, nu = frequency, λ = wavelength, T = temperature.

Respectively, c, h, G and k are the speed of light, Planck's constant, the universal gravitational constant and the Boltzmann constant. These are the conversion factors between space and time, energy and mass, energy and space, energy and time and energy and temperature.

Planck's mass, which marks the threshold of quantum gravity, the Eldorado Physics, is easily obtained by equaling the radius of a black hole of mass m with its Compton wavelength, thus equaling a gravitational length to a quantum length, namely, $2Gm/c^2 = h/mc$. We deduce the Planck mass, and the eponymous length, density, time and temperature, i.e. 10^{19} GeV, 10^{-33} cm, 10^{94} g/cm^3, 10^{-43} seconds, which mark the border of what we can comment on. A conceptual catastrophe occurs when crossing Planck's epistemological wall. Time is foaming. We can no longer talk. We have lived happily ever since matter spoke. The era of lunacy has arrived, a mark of the madness of a mind pushed to its limits and not of this world. Let us therefore close this sad parenthesis.

The time-energy uncertainty allows for an even greater miracle than the seeds of galaxies: the haphazard creation of multiple cosmos.

---

31 Not to be confused with the absolute origin of everything, or if you prefer, the universe.

Hallelujah! We live in a cosmic champagne flute in which we only occupy a bubble. Champagne for everyone!

Curses! The paths to the origins are cut off. We do not know what bubble we are in. We have lost the bubble.

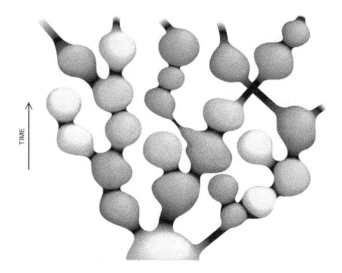

Figure 6.14. *Eternal Inflation: a universe can hide or cover another. It is cut off from the paths to the origins. For a color version of this figure, see www.iste.co.uk/gaudin/chance.zip*

6.7. Landscapes

Just like Jonathan the Heraclitean[32] biologist we are searching for our green vital valley V^3.

The cosmologist of quantum obedience, as well as the biologist, is no more afraid of big numbers than a surgeon is afraid of blood. Likewise, they do not avert their minds' eye. Their advice is to consider that the energy valleys at the top are not isolated and that they can correspond with each other, not by mail, but through the tunnel effect (from below), or quantum gliding (from above), which uses the ascending winds of uncertainty to rise above the mountains – to fly or to dig, given the quantum uncertainty, amounts to the same thing.

---

32 Jonathan Weizman, landscape gardener in biology, and dear friend in heaven. He looked for the source within dream landscapes.

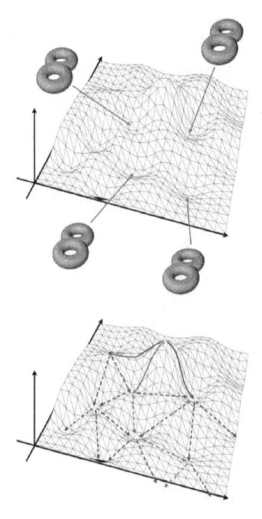

Figure 6.15. *Superstring theory landscape. Thanks to the tunnel effect (mountain pass), it is not necessary to go through a pass to get from one valley to another. It is advisable to not get lost in theoretical swampland so as to not dirty your boots[33]. For a color version of this figure, see www.iste.co.uk/gaudin/chance.zip*

A sudden gust of wind blows a universe into the heavens, allowing it to leap over mountains and pass from one valley to another. We panic: the cosmos are constantly being created and moreover they are infiltrating each other. There is no longer a

33 For further reading on this subject, visit the various articles on the ArXiv physics server.

simple probability to calculate, or even a realistic presumption. We can only reason haphazardly, on the cusp of equations.

The cosmos and life are not mutually exclusive, but raise probabilistic questions.

In the case of the cosmos, we can easily become confused, since both the number of favorable cases (viable universes, potentially holding stars, and life) and the total number of cases border on infinity or infinity/infinity = indeterminate. A new statistical discipline is yet to be born[34]. But for now, let us take it all as metacosmological logos, with a pinch of sky.

6.8. The good works of Inflation

It heals the mistakes caused by the BB[35] and solves the problems of cosmology families, flatness of space, horizon (causality) and laces (division into galaxies). Others are able to talk about it far better than I am able to[36], but let me touch on it lightly.

Inflation stretches space until it removes imperfections, then it turns it into a festoon and embroiders it into a latticework of stars through which the Dark passes.

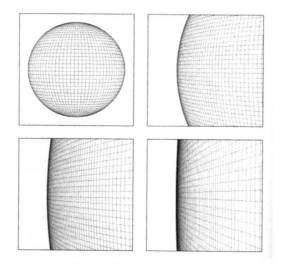

Figure 6.16. *Flattening effect of inflation*

34 Read Vilenkin, Guth and Linde on ArXiv.

35 Big-Bang, phrase used by Hubert Reeves.

36 Jean-Philippe Uzan, *Big-Bang*, Flammarion.

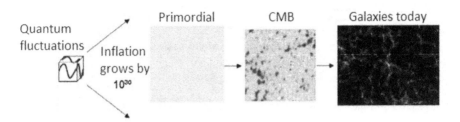

Figure 6.17. *The good work of inflation: origin of large structures. For a color version of this figure, see www.iste.co.uk/gaudin/chance.zip*

Inflation is a space straightener and irons it out, so to speak, taking away its folds. A false vacuum pearl full of repulsive energy grows much larger than the observable universe. The laws of the universe are the laws of the pearl. They are provincial, even parochial.

6.9. Sub species aeternitatis

The superstrings propose, eternal inflation disposes, and the Anthropogenic Principle selects universes good for life. Inflation is eternal[37]. Once it has started, nothing can stop it. Eternal Inflation, combined with the theory of superstrings and the anthropic principle[38], sows the universes in much the same way as others sow wheat, the large, the small, the ephemeral and the durable, the diversity is such that one becomes a landscape painter of the universe.

Quantum is synonymous with fluctuating, its energy needs satisfied by the grace of $E = mc^2$, takes shape. The quantified vacuum becomes a lender and a binder.

In French the anagram of *Vide* (Vacuum) is DIEV; if one reads the 'V' as Latin for 'U' it reads Die, which translates as God; which in physics, would be the best approximation of God if it were not so wildly whimsical. He is the father of space and light. Space gives yourself and space gives itself. The emptiness sheds its energy and lights up: *Fiat lux!*

As Feynman put it, "created and annihilated, created and annihilated – what a waste of time"!

---

37 Not everyone is ready to bestow inflation with this divine attribute, and others, more radical still, go so far as to deny its virtue, and its very existence (see Steinhardt).

38 Received with honors by cosmologists in need of explanation, often criticized by particle physicists.

6.10. The smiling vacuum

The quantum vacuum is not retired, at the micro-scale, it is active and frantic, but perfectly lawful. The conservation of energy and all that is important to the physicist – quantum numbers and various charges – is respected.

Particles are never created if their energy is not returned like empty bottles in uncertain time.

The vacuum is bubbling with activity, but we do not see this because the fluctuations contradict each other perfectly, and, on average, compensate for each other. As we are naive, we see nothing but fire. It is no less than the ultimate peace and serenity that unfolds in the night sky each night, caressing the eye and offering reassurance; the permanent place in which dramas and births unfold in the invisible. It is like the Sun, and the atom, even though they carry nuclear hell in their hearts, their faces are serene. In short, under the luminous skin of objects we can guess there are storms.

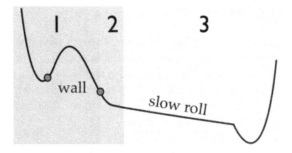

Figure 6.18. *Cosmological landscape in section. Recipe for the universe: Take a field (scalar): (1) tunnel (between the red dots); (2) birth of the bubble-universe; (3) inflation in the bubble and the creation of matter. Bake long enough to solve the problems of cosmology: flatness and horizon. For a color version of this figure, see www.iste.co.uk/gaudin/chance.zip*

May this metaphorical discourse not abuse you or mask the philosophical and cultural breadth of the new and pluriversal cosmogony: in the institutes of astrophysics, the Big Bang and the apocalypse scenarios (death of the proton, big crunch), and perhaps modern mythologies and geneses.

One only needs to allow physics to boil the blood of the species.

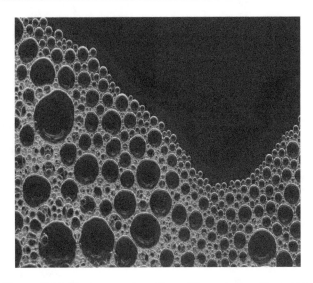

Figure 6.19. *Spumescent amities of the vacuum. For a color version of this figure, see www.iste.co.uk/gaudin/chance.zip*

7

The Chance in Decision: When Neurons Flip a Coin

Imagine yourself as a microeconomist: you are trying to predict the behavior of an agent who has several options in front of him... which one will he choose? Fortunately, some of your older colleagues have looked into the matter. At the turn of the last century (in 1944), von Neumann and Morgenstern developed the so-called *classical* theory of expected utility. The original idea is that making a rational choice amounts to maximizing a function that attributes a utility to each option; in other words, to choose well is to select the option of *greatest expected utility*. The word "expected" refers to the notion of mathematical expectation, as conceived by Blaise Pascal in the 1650s. Indeed, we do not know the utility of each option in advance. We can only imagine the possible outcomes if it is "chosen", and then estimate, for each outcome, its probability and value (the amount of well-being that it would bring). The right choice thus corresponds to the option that maximizes the product of value and probability, summed up over all possible outcomes.

7.1. A very subjective utility

From the observer's point of view, the problem is that we do not know the probabilities or the values that the agent assigns to each of the outcomes. Even in a simple case, like a lottery, that is, a monetary amount associated with a probability, say, a one in two chance of winning €100, people do not follow the objective magnitudes. For example, most people would rather receive €40 upfront than a coin toss to maybe win €100, even if the mathematical expectation (€50) is higher. This is called risk aversion. We can account for risk aversion through a concave

Chapter written by Mathias PESSIGLIONE.

(subjective) utility curve with respect to the (objective) amount in euros, so that the utility of €40 is greater than half of the utility from €100.

This Law of Diminishing Marginal Utility, according to which utility increases less and less steeply as the gains accumulate, was proposed by Daniel Bernouilli in 1713, as a solution to the so-called St. Petersburg paradox. The problem was to explain why people quit playing after a certain number of turns won (heads or tails) in a game of "double or nothing", despite the mathematical expectation being infinite if they adopt the strategy of choosing "double" on each turn. The solution, again, is found through a concave utility curve: after earning a certain amount of money, the utility attributed to "keeping your earnings" is worth more than half of the utility attributed to "doubling your earnings". In classical versions of the theory, the concave curve is obtained by adding a power parameter of less than 1 to the objective gain in the utility function. This parameter is on average around 0.8, but varies greatly from one person to the next – some individuals even have a propensity to take high risks, which corresponds to an exponent greater than 1.

The probabilities are not exempt from subjectivity either. Kahneman and Tversky have shown, in their prospect theory, that there is a tendency to overestimate small probabilities (such as winning the lottery) and underestimate high probabilities (such as not winning the lottery). Here, again, we can account for this subjectivity by a distortion of the probabilities in the utility function. Besides, Kahneman and Tversky have shown that the same options are not valued in the same way, depending on whether they are presented as a gain or as a loss. For example, most people would rather flip a coin for a €100 out of their pocket than agree to give €40 directly. While they exhibit risk aversion in the domain of gains, they are, on the contrary, risk-seeking in the domain of losses. This can be modeled by an inversion of the concavity in the subjective utility curve, between the domain of gains and that of losses.

In addition, when people choose between options that present both the probabilities of winning and losing certain amounts, for example, 90% chance of winning €10 and 10% chance of losing €100, we observe that the possible losses carry more weight than the possible gains. Therefore, Kahneman and Tversky added another parameter into the utility function, which allows for a change of the slope at point 0 that separates the gain and loss domains. This parameter is also highly variable from person to person, with an average that is close to 2. This means that for the average person, losing €100 is twice as painful as winning €100 is pleasurable.

7.2. A minimum rationality

All this is not very rational, you will say. In reality, we can see the successive distortions that economists have inflicted on utility functions, under the pressure of behavioral observations in choice tests, as maneuvers aimed at saving a minimal notion of rationality. This minimal rationality tolerates any utility function, no matter how crazy. We may well be masochists, and prefer hell to heaven, or furiously nationalistic, and sacrifice our lives for the Homeland. In this context, to be rational means to maximize well-being defined by a subjective utility function. In other words, to be rational is to be predictable, once the utility function is known.

The problem, of course, is that the utility function is not directly observable. If we infer it from the choices observed in an agent, the reasoning becomes circular: we are going to reconstruct a utility function such that the choices appear rational *after the fact*. This is why von Neumann and Morgenstern set out to define criteria for rationality, which refer to choices and not to utilities. They show that respecting a utility function amounts to respecting a small number of axioms that concern the preference relation between the options of the choice. For example, according to the axiom of comparability, I can always express a preference between two options A and B: either I prefer A, or I prefer B, or I am indifferent. According to the axiom of transitivity, if I prefer A to B and B to C, then I must prefer A to C. According to the axiom of independence, if I prefer A to B, then I must prefer A + C to B + C, etc.

Rationality therefore does not refer to maximizing profit, as a naive vision of decision-making in economics requires, but to the consistency and stability of preferences. However, even these relatively flexible axioms are frequently violated by real humans in real life, who do not behave like the *Homo economicus* of theory. Depending on the moment and the context, the preferences expressed will change, while the options remain the same.

7.3. There is noise in the choices

There are two types of deviations from the axioms of rationality: systematic deviations (biases) and stochastic deviations (chance). These two sources of instability in preferences complicate the task of the microeconomist, namely, to predict the choices of human agents. While biases can be incorporated into the prediction (once they are spotted), chance appears as an irreducible source of unpredictability.

Let us take an example which we studied in the laboratory: the choice between a safe option and a risky option. The safe option is a small win every time. The risky option can be either a lottery (analogous to a wheel of fortune) or a motor challenge (squeezing a handgrip with just the right amount of force). This risky option is characterized by three attributes that vary across trials: the gain obtained in the event of success, the loss inflicted in the event of failure and the probability of winning (indicated by the width of the target window into which the needle must fall).

This type of choice is precisely the classic case modeled in expected utility theory. The utility is given by an equation of the type:

$$U(gamble) = Proba(gain) \times Value(gain) - Proba(loss) \times Value(loss)$$

which can be rewritten as:

$$U(gamble) = P(gain) \times gain^{\alpha} - \lambda . (1 - P(gain)) \times (- loss)^{\alpha}$$

The parameters of this equation are the concavity (α) and the relative weight of the losses compared to the gains (λ). In order to best reproduce the choices of a given participant, we can adjust the parameters of this utility function, which will become subjective (personalized, in a way). This is called "fitting" the model, through a *softmax* function, which takes utility as the input and generates a probability of choice as the output. In the case of a binary choice, this reduces to:

$$P(gamble) = \frac{1}{1 + e^{-(\tau(U(gamble) - U(certain)))}}$$

We see that with this function, probability is worth 0.5 when the options have the same utility, 1 when the utility of the gamble tends to infinity and 0 when the utility of the certain option tends to 0. Between the two asymptotes, it takes the form of a sigmoid curve, the slope of which is adjusted by the parameter τ. This parameter, sometimes called the inverse temperature, by analogy with Boltzmann's equations, captures the stochasticity of choices. If it is 0, the choices are random: the probability will be 0.5 (one chance in two) regardless of the utilities. The more it increases, the more the choice is determined by the difference in utility between the two options. Fitting the decision-making model, which includes the utility function and the softmax function, therefore amounts to fitting three parameters: α, λ and τ.

There are several ways to achieve the fit. In all cases, it is a question of matching, as closely as possible, the probabilities generated by the model and the frequency of the choices observed. Through this model, choices are thus only predicted on average. In fact, a choice made by a given subject in a given trial can only be 0 or 1 (one option or the other). Except for a trivial choice, where the difference in utility is such that it dominates stochasticity, the softmax function will be wrong in some instances. However, it can indicate the correct frequency (of bets, for example) if we consider a set of choices, observed in one or more individuals.

The stochasticity parameter in the softmax function therefore testifies to a certain inability to predict a single choice, even when the utility function is known. However, it gets worse: the parameters of the utility function are not themselves stable; they vary depending on the state of the agent and its environment.

7.4. On the volatility of parameters

It is a fact of life that people play the lottery more when the weather is nice (especially if it is unexpected), or after their favorite team wins (especially if it was unexpected). This phenomenon is generally interpreted as reflecting the impact of mood on the attitude towards risk, and therefore on the decision to take a gamble. In short, we are more optimistic when we are in a good mood. This is a problem for expected utility theory, because the preference between the same two options (buying or not buying a lotto ticket) will change depending on the mood of the moment. This is a perfectly irrational bias, since the lottery draw is independent of weather and sports results. However, even if this mood bias is known, and even if we can take into account the weather and sports results, we cannot fully correct for it, since most of the other factors likely to make a person be in a good or bad mood are not accessible (even for Google).

Fluctuations in mood, and their impact on decisions, can, to some extent, be reproduced in the laboratory. In one of our experiments (Vinckier *et al.* 2018), we asked participants questions on general knowledge. To each question, they had to give one of several possible answers, depending on which they received feedback ("you win!" or "you lose!"). From time to time, they were asked to rate their mood on an analog scale, graded from "very bad" to "very good". These mood ratings can be modeled as a weighted sum of the feedback received, with the most recent having more impact than the oldest. The model is actually a little more complex, since the causality goes in both directions: feedback affects mood, which, in turn, modulates the perception of feedback (we are less sensitive to negative feedback when we are in a good mood). With this model, we can reconstruct a theoretical mood, based on the feedback history, even if we never ask participants to express their mood.

Now comes the crucial test: between the rounds of the general knowledge test (including the question, the answer and the feedback), we replaced the mood ratings with the choices we talked about above, between a certain option and a risky option. Does the mood model reduce the gap between the choices observed and those predicted by the expected utility function (with the adjusted parameters)? The results confirm that mood influences choice, in that participants were more likely to gamble when their (theoretical) mood was better. In fact, mood impacts the relative weight of gains and losses in the utility function: when you are in a good mood, you take less account of potential failures. Far from representing immutable personality traits, the parameters therefore vary according to the internal state of the individual themselves depending on the context (here, after positive or negative feedback concerning responses to the general knowledge test).

If we do not know the psychological state of the agent (especially their mood), the choices will seem more stochastic. Part of randomness therefore comes about through a lack of knowledge of the factors that will bias the decision. This interaction between the psychological state and the decision-making process necessarily involves brain mechanisms, knowledge of which makes it possible to further reduce the degree of randomness in choices.

7.5. When the brain wears rose-tinted glasses

To further elucidate on these mechanisms, we asked the volunteers to perform the general knowledge test, coupled with the choice whether to gamble or not, in a Magnetic Resonance Imaging (MRI) machine that measures the hemodynamic signal (indicating variations in blood flow) in the different regions of the brain. The theoretical mood reconstructed by the model, which fluctuated according to the feedback received, was correlated with the hemodynamic activity of a particular brain region, the orbitofrontal cortex. However, this region (called this because it is located at the front, just above the orbits) is known to signal the utility of options when making a decision (see Figure 7.1).

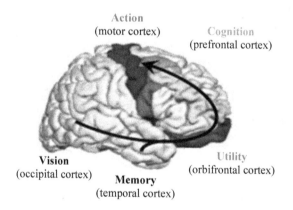

Figure 7.1. *Decision-making in the brain. For a color version of this figure, see www.iste.co.uk/gaudin/chance.zip*

COMMENT ON FIGURE 7.1.– *The options to choose from are represented either in the occipital cortex, at the back of the brain, if they are visible in the immediate environment (chocolate cake), or in the memory systems of the temporal cortex, if they are formed by imagination (vacation in Tahiti). In either case, the utilities of these options are estimated by the orbitofrontal cortex at the front of the brain. These anticipated utilities influence both the systems responsible for action planning, in the prefrontal cortex, and those for action execution, in the motor cortex.*

By testing the impact of basal activity recorded in the orbitofrontal cortex (before options are presented) on utility function parameters, it has been observed that when mood is elevated, the brain gives a greater relative weight to gains over losses. Everything happens as if we see the glass half full: we take more account of the prospects of gain or success – in short, seeing the world through rose-tinted glasses. Knowing the state of the brain system signaling values (including the orbitofrontal cortex) therefore makes it possible to further reduce the (apparent) stochastic nature of decisions.

In fact, symmetrical to the orbitofrontal cortex, in another region of the brain, is the insular cortex, whose activity increases when the mood is lower. To characterize the reciprocal influence of the two regions on decision-making, we can rewrite the utility function with two separate weights on the terms of gains and losses (see Figure 7.2). In this case, we must remove the stochasticity parameter in the softmax function to preserve the identifiability of the model. We can thus distinguish the contribution of the basal activity in the two regions, which reflect the good mood and the bad mood: that of the insular cortex plays on the weight of the losses, while that of the orbitofrontal cortex plays on the weight of the gains. Therefore, there are opponent systems in the brain, which, respectively, magnify the prospects for failure and success. The relative influence of the two systems helps to explain the framing effects, i.e. the fact that the decision changes depending on whether the choice is presented in terms of gains or losses. However, for the sake of simplicity, in the following I will consider a single utility signal, from the orbitofrontal cortex.

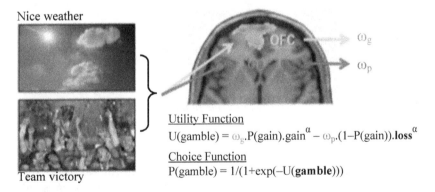

Nice weather

Team victory

Utility Function

$U(\text{gamble}) = \omega_g.P(\text{gain}).\text{gain}^\alpha - \omega_p.(1-P(\text{gain})).\textbf{loss}^\alpha$

Choice Function

$P(\text{gamble}) = 1/(1+\exp(-U(\textbf{gamble})))$

Figure 7.2. *An example of mood bias. For a color version of this figure, see www.iste.co.uk/gaudin/chance.zip*

COMMENT ON FIGURE 7.2. – *Positive events, whether sporting or weather, put the brain in a good mood, which corresponds to an increase in the orbitofrontal cortex basal activity, which, in turn, translates into a greater weight attributed to the expected gains in the utility function that determines the choice whether to gamble or not. The symmetrical mechanism, in case of defeat or bad weather, involves higher basal activity in the insular cortex, which increases the weight of potential losses in the utility function. For the sake of simplicity, we have zeroed the utility of the other options in the softmax function that generates the probability of gambling (of taking a risk).*

7.6. The neurons that take a vote

Even if knowing the bias improves the decision-making model, it is far from perfect: there remains a part that is unpredictable. Instead of introducing stochasticity into the decision-making model, we can track down chance to the brain mechanism that underlies the decision. This mechanism is not fully elucidated, but there is consensus on some models, because they explain several properties of choices, neurally as well as behaviorally. Among these are attractor networks (see Figure 7.3).

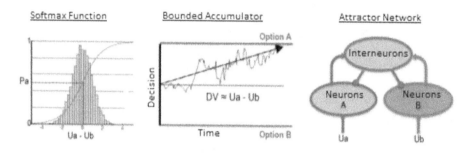

Figure 7.3. *Chance in decision-making models. For a color version of this figure, see www.iste.co.uk/gaudin/chance.zip*

COMMENT ON FIGURE 7.3. – *The softmax function transforms the difference in utility between two options, A and B, into the probability of choosing one or the other. At the indifference point, when both options have the same utility, the probability is at the level of chance (0.5). Otherwise, the degree of stochasticity is adjusted by the slope of the function (often referred to as the inverse temperature). If the sigmoid looks flat, the choices are dominated by chance; if it resembles a step, the choice is dominated by the difference in utility.*

In the bounded accumulator, a decision variable drifts over time until it reaches one of the two bounds corresponding to the two possible options, and thereby triggers the decision. The average of the drift corresponds to the difference in utility between the options, to which a random noise is added at each time step. This model preserves the properties of the softmax function with regard to choice probabilities, and accounts for the distribution over decision times, which are longer for closer utilities.

The attractor network makes the same predictions as the bounded accumulator at the behavioral level (choice probability and decision time), and shows how the choice process can be implemented in a neural network. In this network, two populations of neurons represent the utility of their preferred option, and compete through inhibitory neurons. Network activity converges on one of the two attractors, where one population reaches a threshold level while the other is silenced. The brain systems downstream of the network need only read the activity of the two populations to know which option was chosen. Stochasticity is introduced in the form of Gaussian noise in the input/output functions of neurons. In addition, the input signals (utilities) may themselves have random fluctuations.

For a binary choice, such as the one at hand, the network is made up of two populations of neurons, one voting for the risky option and the other for the certain option (Wang 2002). Each population receives, as an input, a signal proportional to the utility of its preferred option, and outputs a signal (the vote) to a set of inhibitory neurons connected to both populations. Therefore, the two populations are placed in competition, until one wins and silences the other, according to a mechanism that is sometimes described as *winner-takes-all*. It is the winning population that signals the final decision to the whole brain.

The model includes randomness (noise) at several levels. First, there is the utility signal at the input of the network, which is not stable: it presents fluctuations that we present as random, but which could have the function of indicating an uncertainty in the estimated utility. Next there are the input/output transformations operated by the neurons, which include Gaussian noise. In this way, the network will generate decisions that vary from one simulation to another, with a prediction that can be good, on average, as it is for the softmax function. Indeed, the network will more often converge on the attractor (the victory of the population) corresponding to the option of greatest utility. It also has a dynamic aspect, which makes it possible to account, not only for the decision, but also for its duration. Indeed, the network will take longer to converge if the decision is difficult, that is to say, if the utilities of the options are close.

The operation of this network can be seen as carrying out a sequential sampling of the value differential between the two options. If, at each time step, we note the difference in activity between the two populations, we have a decision variable that drifts toward one of the two bounds, corresponding to the choices of the risky option and the certain option. The network is thus close to a bounded accumulator, the decision variable which has, at each time step, a drift proportional to the utility differential, to which Gaussian noise is added. These dynamics of accumulation that shape the decision were effectively recorded at the level of certain neuron populations by means of electrophysiological recording techniques (Gold and Shadlen 2007).

7.7. The will to move an index finger

Where does this randomness in neural networks come from? Most neuroscience researchers see the brain as a deterministic system. There is therefore no generator of chance in the metaphysical sense, but only a limited understanding of brain mechanisms; the chance introduced into the models therefore represents epistemic uncertainty. To paraphrase Laplace: if we had complete information about both brain states and the laws that govern transitions between states, we could predict decision-making perfectly.

Does this then mean that we are not free? This is a question that has been examined by philosophers since philosophy began and has attracted very diverse answers. For Spinoza, for example, men believe they are free because they ignore the causes that determine them. For Descartes, on the contrary, both preferences and moral attitudes can be decided freely. On this point, of course, he had predecessors, like the Stoics, and successors, like the existentialists. Neuroscience has been embroiled in this question since Libet's experiments in the early 1980s.

To understand what is at stake in these experiments, we must immediately remove the obstacles to free will that no one thinks of denying, i.e. those which arise from constraints imposed on the subject, for example, the fact that one cannot fly by flapping one's arms, or kill one's neighbor without risking prison. Libet is interested in a decision that does not involve preferences or morals that can be influenced by external factors, from genetic background to the social environment. In fact, he is interested in a decision that has no interest: that of moving (or not moving) his index finger.

The seminal experiment involves asking participants to press a button with their index finger, when they feel like it. A clock is placed in front of them with a hand that goes round quickly (in 3 sec). The participants must, thanks to this clock, report the time at which they become aware of their intention to move. At the same time, the potentials generated at the level of the premotor cortex are recorded, using electrodes placed on the top of the skull. There is a potential at this level, related to motor preparation, which rises linearly in the second that precedes a movement of the hand.

The result is based on the timing of events, when locked onto the time of pressing the button with the index finger. Participants typically report realizing their intention to move about 200 msec before the movement, while the potential for motor readiness began to climb long before (about 600 msec earlier). So it seems that when one thinks of exercising one's conscious will (to move one's finger), the brain has already made its decision – which Libet presents as evidence refuting free will. In fact, everything happens as if the experimenter, who observes the brain potentials, could predict, before the agent, the moment when the latter will move.

7.8. Free will in debate

This experiment has caused a lot of ink to flow; it is impossible to reproduce the debate here *in extenso*. I would just like to highlight a few points for reflection. The first is technical. Some criticism has focused on the fact that spotting the position of the hand on the clock also takes time, which may explain the delay of conscious perception over brain potential. This objection was swept aside by later demonstrations, showing that it is possible to predict movement a lot longer in advance (up to several seconds), from other brain signals, measured by functional MRI, or by electrodes implanted in the brains of patients for medical reasons.

Other, more subtle critics, have noticed that the counterfactual case (when there is no movement) is never taken into account, since the electrophysiological recordings are observed only during the second preceding the movement. It is therefore possible that the potential may also rise in the absence of movement. In this case, it would not sign the decision to move, but simply a random fluctuation. Indeed, we can reproduce the phenomenon by simulating the decision to move with a variable, which drifts and triggers the movement when it reaches an upper bound (Schurger *et al.* 2012). If we average the trajectories of this variable, by aligning them with the time of the movement, we obtain a signal that rises in a linear fashion, like the motor readiness potential. However, this ramp signal in no way indicates that the decision has been made, because, by construction, the decision in the simulation is only made if the variable reaches the upper bound.

This criticism is relevant to Libet's original experiment, but it does not apply to more recent experiments that have been done to predict, from brain signals, not the time of the movement, but the choice – for example, that of turning left or right at the next junction in a driving simulator (Fried *et al.* 2017). However, there are other problematic points, for example, the possibility that the participant does not really report the moment of conscious intention, but instead when the decision seems irreversible. In this case, as long as the brain signals indicate the decision, it would have been prepared well in advance, but the participant would retain the power to block or reverse it at the last moment. This conception thus maintains a notion of free will in the power to say no, even once a decision has already been initiated.

It should be added that the situations studied in the Libet experiments are very particular, because participants are asked to make a decision freely, but repeatedly, for options that remain identical. It is perfectly possible for participants to decide ahead of time what their next decision will be, and in this case, it is not surprising that this decision can be read into their brain signals. Moreover, the decision-making observed in these situations, where there is no reason to choose one option over another, probably does not say much about the decisions that matter, the ones that will change the life of the agent and self-representations.

7.9. The virtue of chance

Without continuing the debate, it seems fair to say that the neurosciences are struggling to formally refute the notion of free will. What they have done convincingly is dissociate the systems that generate the impression of conscious will and the systems that actually make the decision (Desmurget *et al.* 2009). Through appropriate interventions, pathological situations can be generated, where the participant has the impression of having initiated an action, however, it was externally triggered. Conversely, there are pathological situations where the brain can trigger an action without the participant having a sense of agency. The consciousness of exercising free will is therefore probably not causal in the generation of action. However, the two systems, the one that makes the decision and the one that creates the impression of control, usually work together.

The neurosciences would have even more trouble if they intended, on the other hand, to prove the existence of free will. We simply would not know how to conceive (and thus test) the intervention of a mental cause (the will) on decisions (the actions taken). Even if this is mostly implied, neuroscience tends to consider that the mind and the body are two modes of the same substance (as Spinoza would have said) and not two separate substances (as Descartes would have said). In this context, there is no reason to suppose a causal intervention of the mind on the body.

And where is chance in all of this? There is a lot of noise in neural processes, in the sense that the outcome of these processes is not controlled. This is the same notion as the chance at work in the game of heads or tails. If we controlled the initial conditions of the throw (the position and speed of the coin), we could perfectly predict the outcome. Likewise if we controlled the outcome, for example by putting the coin back on tails every time it lands heads, it would be completely predictable.

However, since the observer does not control either the beginning or the end, decision-making remains somewhat unpredictable. Some wanted to see a space for free will in this stochasticity. However, while free will does imply unpredictability, unpredictability does not imply free will. Just because you cannot predict which way the coin will fall does not mean that it has a will of its own.

There is, however, a virtue in the stochasticity of decisions. If we always went with the option of greatest expected utility, we would greatly reduce our exploration space. We would persist in using routines that are certainly valid, but perhaps largely suboptimal, in view of the unexplored possibilities. If natural selection has not favored the evolution of control systems, which would make it possible to reduce noise in decisions, it is perhaps because this noise has an advantage: it allows agents to explore options whose expected utility seems weaker *a priori*, but which *a posteriori* will turn out to be new El Dorados.

7.10. References

Desmurget, M. and Sirigu, A. (2009). A parietal-premotor network for movement intention and motor awareness. *Trends in Cognitive Science*, 13(10), 411–419.

Fried, I., Haggard, P., He, B.J., Schurger, A. (2017). Volition and action in the human brain: Processes, pathologies, and reasons. *Journal of Neuroscience*, 37(45), 10842–10847.

Gold, J.I. and Shadlen, M.N. (2007). The neural basis of decision making. *Annual Review of Neuroscience*, 30, 535–574.

Kahneman, D. and Tversky, A. (1979). Prospect theory: An analysis of decision under risk. *Econometrica*, 47(2), 263–291.

Libet, B., Gleason, C.A., Wright, E.W., Pearl, D.K. (1983). Time of conscious intention to act in relation to onset of cerebral activity (readiness-potential). The unconscious initiation of a freely voluntary act. *Brain*, 106(3), 623–642.

Schurger, A., Sitt, J.D., Dehaene, S. (2012). An accumulator model for spontaneous neural activity prior to self-initiated movement. *Proceedings of the National Academy of Sciences of the USA*, 109(42), E2904–E2913.

Vinckier, F., Rigoux, L., Oudiette, D., Pessiglione, M. (2018). Neuro-computational account of how mood fluctuations arise and affect decision making. *Nature Communications*, 9(1), 1708.

Wang, X.J. (2002). Probabilistic decision making by slow reverberation in cortical circuits. *Neuron*, 36(5), 955–968.

8

To Have a Sense of Life: A Poetic Reconnaissance

Why talk about poetry at a conference entitled *Chance, Calculation and Life* (Cerisy, August 2019), where the main specialties concerned are mathematics, biology and philosophy? I will not do so as a literary specialist, which I am not, but instead as a polymathic nomad for whom disciplinary boundaries are made to be blithely crossed. Hence I will start by asking my scientific friends a direct question: Is there such a thing as a *life* science?

Each science constructs its object. Is life reducible to the object of biology? To the object of one specific science, or of science in general? And if so, "*tout le reste est litterature*" (the remainder is literature)? In any case, this formula is not of a narrow-minded positivist; it is a verse by Paul Verlaine – taken from his poem *Art poétique*, which opens with the no less famous line: "*De la musique avant toute chose*" ("Music before anything else").

In this formula, the word that holds me back is the word *reste* (remainder) – where I hear the *remainder* resulting from a division. The division of mind by and across all disciplines – literature included. And in this irreducible residue, I believe I see the substance of essential knowledge, for an adequate standard of living, to which the term poetry belongs.

My approach is in response to the following hypothesis: if life concerns many "logies" (from biology to theology), for many sciences, including those that deal with chance and calculation, there is an essential *remainder* that endures. A remainder which is not about reducing, but about describing ... not a question of mythologizing or enchanting, but of *reconnaissance*, in every sense of this beautiful

Chapter written by Georges AMAR.

French word (from cognition to gratitude). This involves a process of reacquaintance – *seeing afresh* – and with new appreciation... a poetic reconnaissance. Today, is this not a necessity for the whole scope of our relationship with reality?

*

Poetry, or better still, the poetic mode of being-acting-thinking-feeling can constitute a renewed resource in terms of our understanding of things, of the world, and of ourselves: this intuition, acute or woolly, can today be found among diverse cultural contexts. Can it go beyond the vague; can it go beyond its decorative "cultural" status? This is the challenge...

Is there a poetic way of understanding things, of welcoming them in a form, without shaping them or pulling them out of their world, of entering into a respectful relationship with them, in which we are all involved? If such a mode of knowledge exists, one of its privileged archives is to be found in the *oeuvres* of poets, insofar as we know how to look for it: it is not in its admiration or in its enjoyment that it can be found. What interests us, through their works and sometimes their lives, is rather "that which the poet *knows*" – about life and the world. Even those poets who do not write poetry...

My premise will assume and follow the axiom of the existence of poetic knowledge, adequate for life, and multifaceted. I will proceed in this mode of inquiry, in search of such knowledge as found in texts and documents, in diverse forms sometimes disparate from any literature. In the sample below, I shall simply listen to a few sentences, some verse and try to identify an underlying conceptual pattern.

*

This reflection has two starting points. I borrow the first from a philosophical biologist: "*La vie n'est pas un concept biologique*" (Life is not a biological concept), as said by Henri Atlan[1]. This little phrase had an enlightening effect on me. I hear two epistemological prejudices here. First is the principle of an absolute discontinuity of life with regard to *so-called* inert matter, a concept that is far too metaphysical for science. It does not "need this hypothesis" (as Laplace said about God) to explain the properties of *so-called* living organisms or tissues. The second preclusion targets the uncontrollable polysemy of the word "life" as found in ordinary language, in "everyday life". A double aversion therefore, to the place of

---

1 See the section *Zoon et Bios* (Atlan 2003).

myth and opinion, which is the basis of science and its undeniable fruitfulness[2]. But at what cost? The dissociation of science, meaning and speech – and its divorce from "public intelligence"? (Stengers and Drumm 2013).

The second starting point of my reflection is the adage of poet Kenneth White: "The poet is the one who has a sense of life". Where did I hear this phrase in his vast work? (White 1987, 2017). Did I half make it up? If it belongs to him, I will borrow it from him, or else I will lend it to him! Moreover, to quote, that is to say to change the context, is already to interpret, to recreate. Behind its apparent simplicity, it says something unique. The poet does not *know* the meaning of life – if it has one. The poet has a *sense* of life, as some have a sense of friendship, rhythm, humor ... or business, wherein the word "sense" denotes a form of intelligence, accuracy... a form of art. But why the poet? And what *life* is it about? And what kind of *art*? Let us continue listening to verse.

The following is from a poet of their own genre, in actions and in things more than in words, although his linguistic prowess is far from negligible[3]. Not content with having invented the "ready-made", the word and the thing, he has sometimes given himself an additional identity in the guise of a woman, bestowed with an unusual pseudonym of his own invention: *Rrose Sélavy*. That the 20th century's most open-minded artist chose this most trivial expression, "*c'est la vie*" (that's life), is telling of itself. Equidistant from the wonders of science and the prestige of metaphysics, which we say without ceremony "*c'est la vie*", is worthy of art. Let me clarify here that Rrose Sélavy's other name is Marcel Duchamp. The very one, whose biographer, Henri-Pierre Roché[4] affirmed that his most beautiful work was his schedule, who, refusing to being called an artist, would eagerly respond to any who asked: "What are you then?" – "A *respirator*, nothing more!"

What of the relationship between the poet and life? Let us continue observing it through the works of poet and philosopher Michel Deguy, in particular in his book which bears the beautiful title of *La vie subite* (*The Sudden Life*)[5]. And whose subtitle sounds like the statement of a theory: *Poèmes Biographèmes Théorèmes* (Poems Biographemes Theorems). The inference here, the accord, or at least the consonance between the registers of art, existence and science, is that the poet is the guarantor – the "undivided poet" invoked by Saint-John Perse in his Nobel

---

2 One of the themes of Atlan's work is the "Intercritique of Science and Myth" (Atlan 1986, 2018).

3 See Duchamp (2008), texts and pieces of his work from 1910 to 1950.

4 (Roché 2012).

5 (Deguy 2016).

speech of 1960. This subtitle has the merit of rhyming "poem" and "theorem", but further still, to introduce between them the third term "biographeme" (a tribute to Roland Barthes). There is a deep connection between life and writing, *bios* and *graphein*. We note in passing that several disciplines have distinct, yet coherent or compatible, *logical* and *graphical* components – I am thinking here of *ethnology* and *ethnography*, or that of *geology* and *geography*. Life is not that lucky in our culture. *Biology* and *biography* are not even estranged sisters... Our next author is still a poet, although he wrote few verses and is read and appreciated for something else. Henry David Thoreau, whose book *Walden; or Life in the Woods* (first published 1854), is enjoying a new lease of life since its growing popularity as a precursor to pragmatic and radical ecology. His other well-read work is that of his *Essay on Civil Disobediance* (1849), which inspired the likes of Gandhi. The two quotations, which we will examine below, come from the monumental *Journal*, which is his "matrix" opus.

"How to make the getting our living poetic! for if it is not poetic, it is not life but death that we get."[6] Along this line of inquiry, the relationship between life and poetry takes on a dramatic tone when considering the allures of political and economic acclaim (the Industrial Revolution has begun). For Thoreau, the Poet is the one who knows how to "make a living", by also working with their hands, with the condition of not "losing their life to earn it". Why does he call this disalienation poetic? Does this seem distant from academic literature, for instance biology? In any case, writing and reading (and lecturing, at which he excels) are at the center of his action and his activism. In addition, Thoreau demonstrated throughout his life a pronounced interest in the scientific process. His precise and exacting observations on the climate, flora and fauna of Walden Pond earned him a reputation as a zoo botanist. He even feared that his taste for science would dampen his poetry, his love and feelings for nature and life. "Suck out all the marrow of life" was his motto and his work. Reminiscent of the "*substantifique moelle*" (substantial marrow) that so obsessed Rabelais! One of the characteristics of poets is certainly the inseparability of flavor and knowledge (*saveur* and *savoir*). May their *science of life* long thwart the dissociation of knowledge, enjoyment and existence (including economic and political aspects).

The second quote from Thoreau's Journal directly addresses the question of biography: "He is the richest who has most use for nature as raw material of tropes and symbols with which to describe his life". What a beautiful alliance between *knowledge of the world* and *knowledge of oneself*. This is the richness of the poet to know how to find (invent) in the things of the world the "tropes and symbols" of an "auto-eco-bio-graphy". Write your life – through the world. And if Thoreau does

---

6 Journal II, Chapter III, JANUARY-APRIL, 1851 (ÆT. 33) (Thoreau 2011).

not really succeed in bridging the gap between *biology* and *biography*, by coming up against the inevitability of their divorce, at least he indicates that it is through the lens of poetry that such dissociation can be rethought, if not repaired.

Let us not overuse quotes. My last is by Rimbaud, who writes and exclaims, in *Saison en enfer* (*Season in Hell*): "True life is elsewhere. We are not in the world". What *life* is he talking about? Neither molecular nor celestial, but "of *that* world", the one to which we refer when we say "*c'est la vie*" or "*c'est pas une vie*" (that's no life). Real life, the poet claims on behalf of all. Perhaps he suffers more or "better" than the others from its absence. Perhaps he knows something about life that no one else sees. We remember the sentence that Rimbaud had Verlaine say about himself in *Saison*: "Perhaps he has some secrets for changing life? No, I would say to myself he is only looking for them". Nevertheless, to seek or to find, for the poet it is the same thing. He knows real life, and he knows the fullness of being in the world, even through its absence.

Finally, according to Hölderlin, "to live is to defend a form". It is a form of life, which is singular and common. We understand that a poet understands this so very well. That he has this *sense* of life.

*

In order to further understand this sense, let us question the nature of the exercise in the manner through which we have approached it: the citation. Where does the strange status we give to certain phrases come from? Why do we accept truth in a popularized expression without any further verification protocols, as science claims for its part as necessary? Without demonstration or argument, as mathematics and philosophy require? One would say that the poem is a *theorem without precondition*, a judgment with no palaver! Why do we listen to Rimbaud, Hölderlin, Thoreau, White or Bonnefoy? For instance, the latter writes, is there a "*vérité de parole*" (truth of speech) (Bonnefoy 1988)? Cézanne wrote in a letter to Émile Bernard, "I owe you the truth in painting"[7]. What kind of *truth*? Truth not *on* painting but *in* painting (speech, music, etc.). On what then? Not *on* life, but *from* – life – heard, painted, spoken.

We are therefore brought back to the hypothesis of the knowledge of poets – painters, musicians and many others beyond the noble arts. The truthful force of certain expressions, including musical or pictorial ones (or even of *phrases* beyond language), what kind of knowledge does this come from? A knowledge that is not about life but rather *of* life listened, painted, said.

---

7 (Derrida 1978).

An intelligence *of* life. Echoing our starting point – the poet's sense of life – this expression makes it clear: the intelligence that life *has*. Life is not only an *object* of knowledge and of science (biological, ethological, etc.). Fundamentally, it seems obvious: life thinks, "knows". Life is intelligent.

However, this axiom, with its ethics, not to make life an object (nor conversely a mystery); only a poet can *grasp* this. The intelligence of life – the mother of all forms of partial derivative intelligence – only a form that has remained faithful to its integrity may echo it, update it, without betraying or truncating its essence. In our civilization, which has for so long consented to the disjunction of faculties and disciplines, which acquired hyper-performance at the cost of hyper-specialization, there is little else other poetry that can achieve this. Poetry still remembers *undivided intelligence*. It knows it intrinsically, *technically* one might say. Poetry is inseparably knowledge and doing. To borrow the Old Latin, it is an *ars*, a sense combining, arranging and tying. The art of composition: to know and to do; the ambit of "content" and "form", etc.

But poetic virtue is better than a conciliator of the opposites, or dialectic in the worn-out forms alluded to above (the debris of yin and yang). Beyond the usual figures of harmony or the happy medium, of tension or the "balance of power", poetic virtue preserves and exercises the intact spectrum of integral *intellectus* – as illustrated by this "chromatic"[8] trio of qualities that are characteristic to all poetry: *acuity of mind, freedom of movement, attachment to the world*. To be poetic is to know how to "grasp life" according to several (that is to say, three) simultaneous, orthogonal perspectives: (1) in its affirmative, irreducible, singular essence; (2) in its games, its dances, its mutations; (3) in its empirical diversity, according to environments and climates.

*

The first thing life knows how to do is survive. This is called the survival instinct; as if this sophisticated ability had to be restricted to a series of reflex actions. Surviving is itself a great art. A very general *ars*, similar to an essential notion from Spinoza's thinking: *Conatus*, which he defines as "the effort by which everything tends to persevere in its being"[9]. All things, and not just living things, which also tend to increase their power to *be-act-think-feel*. The same definition adds that *Conatus* is "nothing but the actual essence of the thing."

---

8 In reference to the chromatic spectrum: three colors (red, green and blue) are enough to generate all colors.

9 Ethics III: On the Origin and Nature of the Emotions, Propositions 6 and 7 (Spinoza 1996).

The elementary competence of life is the affirmation of a singular self-being. To this corresponds the first principle of poetic intelligence, the sense of the singular: the ability to see and say things in their singularity, and to experience their own actuality. Yes, I recognize a voice or a gait among a thousand others; I know, say, the "thrill of reality" (Segalen's "passion for the Real"); the instant (D.H. Lawrence's "quickness of time"); "the virgin, vivacious and beautiful today" (Mallarmé); the singular, which escapes any attempt to shape it, including by identifying it. Every living being perceives this. The living knows the quickness.

The second axis of vital knowledge is that of play. To play, to fool. What could be more playful than a little kid or animal for that matter? Didn't life invent the extraordinary lottery of sexual reproduction, as if the cards had to be reshuffled with each conception? Life loves games of chance. The games of love and chance (Marivaux) or those of chance and necessity (Monod). Life/poetry often likes to contradict itself, cherishing both suitable form and well-regulated games, as much as it relishes breaking the rules and disrupting form, until chaos – which is the supreme form. Life is that which sings and dances, with the madness of Dionysus, as well as with the measure of Apollo. For pleasure. To explore. The possibilities. Even for nothing. To live for livings own sake. Pure song.

The third aspect of the intelligence of life is that of survival. The French, *survivre* (to survive) is derived from *vivre* (to live), so when it is broken down into its two parts, we have *sur* (on or above) and *vivre* (to live). It is not enough for the living not to die, but to push for adventure. To survive and *sur-vivre* (a richer or augmented life), it must open up to something bigger than itself. Every living thing has its *umwelt* (as ethology pioneer Jacob von Uexkull puts it), its own world, which it defines and which defines it; Gilbert Simondon, in reference to technical beings, termed this "associated space". Its "environment" is more vague.

No life without a world. "Small worlds" (the sociocultural circles we are sometimes wont), or "open worlds" – those that "Free man, you will always cherish" (homage to Baudelaire) – the sea. Or this beautiful theorem of Shakespeare's: "There are more things in heaven and earth, Horatio, than are dreamt of in your philosophy"; a line given to Hamlet. Expansion! Breathing! Joseph Delteil, a self-described Paleolithic poet, saying for the little bird: "Piercing, we all have beaks for that"! To open, and to open oneself; expand, embrace, welcome, opt, choose. There is in life – and in poetic intelligence which is this same life accompanied by language – a principle of expansion, which is not the endless accumulation but the enlargement of Self. A taste for adventure that is not prowess but rather a path toward the non-Self. A little dialogue that I once read scribbled on a wall in the Paris metro: "– I want to be an alpinist. What do I need Sir? – First, a mountain." A world!

*

To conclude, before reweaving the threads briefly drawn here, I would like to place my remarks in the current context. If I invoke and outline the traits of a poetic intelligence more faithful to the genius of life, it is not to dispute the value of science, nor is it to oppose a so-called rational approach by another that is closer to myth, or the imagination, or the arts in general. The question of life like that of the planet, at a time of foreseeable technological, climatic, political (and health) upheavals – the horizon of a supposed Anthropocene – calls for a (r)evolution of intelligence, and hence the need to revisit all resources.

Around 1665 Spinoza wrote and left unfinished a *Treatise on the Emendation of Intellect*[10], which foreshadowed *Ethics*. Such a reform ("emendation"), even an overhaul, is periodically necessary. This was the century of Spinoza (and Descartes, Newton, Leibnitz, etc.). It is particularly true for our own times, which have understood that the dawn that is before it (perhaps far away) is not a matter of wiping clean the slate of the past, nor of euphoric progress, nor of fallacious nostalgia; but instead a mobilization of *all* the resources of the human spirit, a re-founding integration of all faculties, individuals and societies, alarmed by the perception of their own imperative and their relationship to the world. And what better name to give to this vast movement than *poetic*?

*

Poetic intelligence/life intelligence is achieved, as we have seen above, by moving along its three distributed axes of triadic inspiration: neither monist, nor dualist (or dialectic); but without rejecting these two essential tendencies of the mind. It seems complex to us, this early poetic intelligence, because it *includes* multiple faculties, knowledge or disciplines that we have become accustomed to dissociating or even opposing. Its presence in our culture is paradoxical. On the one hand, everyone has it. The complexity of its (triadic) analysis does not prevent it from representing the native disposition of every being – living, singular, existing. As soon as it is born, it holds onto itself; it plays, it experiences and it unfolds; it meets and invents a world for itself. These three skills constitute the *sense of life* without which no one can survive for long, much less grow.

---

10 "Traité de la réforme de l'entendement" is the classic French translation of the original, *Tractatus de Intellectus Emendatione*. A recent translation (Pautrat 2005) favors *Traité de l'amendement de l'intellect*. However, I have a soft spot for the old word *"Entendement"*, which to my ear enriches the intellectual aspect (*"entendre"*, to hear) with sensitive and affective notes (tune)...

On the other hand, this primary, *implex* intelligence is dominated and obscured by the reigning separative forms of knowledge and intelligibility: the sciences, the arts. Religions. Disciplines. Of course, it does not disappear, but resists in the background, finding refuge in the vernacular or elementary forms of culture. Poetry remembers this inseparable phase of the mind. Poetry, where the games of meaning and sound, of the "I" and the "world" are always heard, and it lucidly remembers the primary intelligence – acute, playful, open to the world.

The poetry of which I speak here belongs not a particular compartment, precious or savage, to the "literary space", but rather to a regime of cognition and action preceding the divisions of the mind and the *ars* in general – logically, chronologically earlier (*upstream* rather than *before*) – therefore, always present, latent, accessible. The fact remains that "literary poetry" remains one of the most important aspects of our culture, privileged laboratories of *"universal poetry"*.

*

Is it utopian to envision an "*intellectus emendatione*" for the challenges of the present time? On the contrary, is that not the only sensible thing to do? Poetic intelligence will be a resource in the proper sense: the source of the *re*. *Re*visiting our relationship with the world, *re*thinking the "meaning of life", bringing into *play* the genius of Sapiens, on broader and richer bases (in particular those of so-called artificial intelligence).

This reform will have a radical critical dimension. "Critical Poetry" for all disintegrative, specialized, unilateral activities – including poetry and science in their own compartmentalized process, since they also sometimes contribute to the promotion of the "one-dimensional man" (Marcuse, 1964). It can be said that there are enemies more urgent than that of *dissociated* science or art – for example, neoliberalism, nationalism or consumerism. One might think that the political sphere is far more at risk than the poetic. But this very disjunction is part of the problem, and sustains the (multi) one-dimensional, (multi) one-sided perspective, which has been our dominant paradigm for too long, fruitful and destructive at the same time.

In this sense, I allow myself this play on words: the great *artist* of tomorrow will be an *arrêtiste* (stopper). We will need the most creative, daring and subtle artists to *stop*, to suspend, in order to deeply reorient all kinds of processes that have become morbid through the forces of unilateralism.

*

Poetic reform of intelligence will be critical, only in order to open up new explorations of human "doing", its *poiein*. Is it not significant that our language has kept from this ancient Greek verb, *poiein*, only its substantive forms: poem or poetry? It is the result that is favored, the fixed form, sacralizing it, against the "living movement". It dissociates the noble form of "making", which can become art, from the utilitarian form of "doing", which became fabrication. How enchanted I was a few years ago when I discovered, in a village in the Peloponnese, the sign of a baker using *Artopoieio*, the maker (*poietes*) of bread (breadmaker), to describe that beautiful profession.

Poetic intelligence is everywhere, without noise, incognito, minor, outside the walls (academic and others). It insists and persists in artisanal trades, "arts and crafts", the vernacular and popular knowledge. Even beyond the human realm, we might admit the notion of "generalized poetics" as formulated by Roger Caillois (1978), who did not hesitate to apply it to stones.

We can therefore explore a vast *geo-exo-poetic* field. All practical, contextual forms of the effective intelligence of "poets in their own genre" – *exopoets*, a discreet majority, without borders or fixed qualities, as distant from the demagoguery of "all artists" as it is from symmetrical aristocratism. This investigation, this study, is an ambitious explorative project for our time. It is a theoretical, practical and open work, which many of us conduct in various and sometimes contradictory ways.

Seeing and speaking of poetic intelligence *among others* will be the true vocation of poets.

*

Life may be infinitely complex, since each of its parts (the brain, for example) seems to be as much as, or more than, the whole – if there is such a thing. Science has enough material to hit us with *hyperplexity* – potentially throwing us into the arms of "oversimplification", which abounds in the marketplace of faiths and ideologies. Poetic intelligence avoids this double pitfall. Its *triplex* logic welcomes the complication of reality and resolves upon itself to act upon the operadicity of a *sense*.

*

To have a sense of life.

Do we live in a time that, while sacralizing it, has lost the *sense* of life when we need it more than ever, given that power seems to be in the hands of the sorcerer's apprentice? The unheard-of prospect of the possible exploration of artificial or extraterrestrial life again raises the question, "What is life?"[11]. Far beyond the framework of biochemistry, even bioethics, it engages the human sense, the open sense toward life. Perhaps this subtle sense is part of "those things that you know when you're not really thinking about them" that the "moralist" Joseph Joubert spoke of with finesse at the turn of the 18th and 19th centuries? But we may not have the option, nowadays, to not think about it ... Only a poetic approach could "think" on it without major conceptual, philosophical, religious or industrial stumbling blocks. The meaning of life cannot be related to a single principle, or to a dialectic, even a sophisticated one. It is more the ability to think on it without losing sight of it out of the corner of one's eye. More than explaining it, it is a question of remaining faithful to it, of remembering it. How? The *Ars Memoria* or *mnemotechny* (attributed to the Greek poet Simonides of Ceos in the 6th century BCE) might give us an idea how. This "technique" that runs through the centuries is far greater than a commodity, being swept away by computers and information technologies ... *Mnemosyne*, the goddess of memory, is the mother of the nine muses! Cicero was a master of this art, then Giordano Bruno, Raymond Lulle, Shakespeare perhaps: his Globe Theater was inspired by "memory theaters".

The trifoliate scheme that I have proposed here is little more and no less than a "memory aid" on poetic intelligence, on the sense of life.

To have a sense of life is to have a *sense of the essential, a sense of play, and a sense of the world.* (1) A sense of the singular, the lively and the instant. (2) A sense of form, language, energy and dance. (3) A sense of adventure, openness and hospitality. Any apprehension about life, scholarly or existential, thats lacks one or other of these three "senses" (and each is more than one), can only be a hindrance. Life, that of the planet and ours on it, needs to be loved as well as known, and experienced through childlike play! All this – the poet (in each of us) knows.

8.1. References

Atlan, H. (1986). *A tort et à raison*. Le Seuil, Paris.

Atlan, H. (2003). *Les Étincelles de hasard*, Volume 2. Le Seuil, Paris.

Atlan, H. (2018). *Spinoza et la biologie actuelle*. Odile Jacob, Paris.

Bonnefoy, Y. (1988). *La vérité de parole*. Mercure de France, Paris.

Caillois, R. (1978) *Le champ des signes : récurrences dérobées : aperçu sur l'unité et la continuité du monde physique, intellectuel et imaginaire ou premiers éléments d'une poétique généralisée*. Hermann, Paris.

---

11 (Shrodinger 1955).

Deguy, M. (2016). *La vie subite : poèmes, biographèmes, théorèmes.* Éditions Galilée, Paris.

Derrida, J. (1978). *La Vérité en peinture.* Flammarion, Paris.

Duchamp, M. (2008). *Notes.* Flammarion, Paris.

Roché, H. P. (2012). *Correspondance Marcel Duchamp – 1918–1959.* Les Presses du réel, Dijon.

Schrodinger, E. (1955). *What is Life? The Physical Aspect of the Living Cell.* Cambridge University Press, UK.

Spinoza, B. (1996). *Ethics.* Penguin Classics, London.

Stengers, I. and Drumm, T. (2013). *Une autre science est possible.* La Découverte, Paris.

Thoreau, H. D. (2011). *The Journal, 1837–1861.* NYRB Classics, New York.

White, K. (1987). *L'esprit nomade.* Grasset, Paris.

White, K. (2017). *Lettres aux derniers lettrés.* Éditions Isolato, Nancy.

9

<h1 style="text-align: right;">Divine Chance</h1>

9.1. Thinking by chance

We often hear that chance is one of the ways through which God presents Himself to us. Because this sentence is about God, many would choose to think that it only applies to believers, when in fact, it concerns everyone.

When an event occurs, our reflex is to give it a simplistic explanation so as to have an immediate representation of reality. If something is outside that which we had not thought of, and disturbs this representation, we eliminate it. This is the reason why in the context of immediate and simplistic explanations, chance can never exist.

However, human experience is not all about the reduction of reactions. If there is survival, and its simplifications, then a full life exists by virtue of being aware of life. In this full life, there is no question of resorting to a simplistic vision. On the contrary, the less we simplify, the more detail of life we go into, and the more life we process, the more alive we become. The more alive it becomes, the more chance begins to exist.

Inner life means opening one's eyes, ears and sensitivity by gathering as much information as possible on what is happening. As a result, an enriching dynamic occurs across three elements. The first is to turn reality into chance. The second is to transform chance into a need. The third is to transform the external need into an internal one.

To perceive that the world exists and that it is rich, one has to get out of the mundane logic of things. Turning the mundane into the unexpected makes the Real

---

Chapter written by Bertrand VERGELY.

original and forces one to look at it, listen to it and experience it. Subsequently, turning the original into an inner need makes it possible to create, not a simple curiosity, but a teacher that continues to teach others. Finally, to live by this teaching, not only once, but by applying it to everything we do, all the while feeling increasingly called to this teaching life, enriches one's inner life. God can then begin to make sense, and with this comes the idea that He speaks through chance. If there are earthly and human teachings, then there are heavenly teachings that reveal the Unseen world, humanity and existence.

Chance is at the heart of inner life, the spirit and creation, and, in order to develop, one artfully makes it exist. In order to unleash the creativity of the imagination, the Surrealists practiced writing at random whatever came to mind. When they were pleased, these chances were said to be "objective" and decisive discoveries. Shaking up the opposition between chance and need, these chances reveal that there are chances that are not actually chances. These chances have the ability to create, and do not simply come from the unconscious but from elsewhere, from a primordial creative breath of life, giving meaning to the idea that there is a divine chance, which of course calls for further clarification.

9.2. Chance, need: why choose?

Typically, *chance* occurs without any connection to that which defies logic, while *need* is related to that which respects logic. Seen from this perspective, everything opposes chance and need, in appearance at least.

So, let us posit chance as such. When it is pure chance, it is no longer chance but a need. Constantly being what it is, bound to itself, it is not linked to anything. Likewise, let us pose needs as such. When it is a pure need, it is no longer a need but chance itself. Being what it is without our knowing where it came from, linked to nothing, it is pure chance. Hence, the paradox: chance and need are not what we think they are. We think that chance is the opposite of needs, but it is itself a need. Conversely, we think that a need is opposed to chance, but it is itself a need. It is not by chance.

When we are in reflection, chance and necessity are not opposed. Reflection consists of thinking what we think, when we think what we think, and entering into thought, and we are both in thought and beyond it. Since we think what we think, we are inside the thought. However, being in the thought of the thought is not simply being in the thought, one is beyond it. Therefore, when we are in reflective thought, we are simultaneously within interiority and within transcendence. We are within the brilliant and dazzling creative spirit.

All human creation follows a logic. At the outset, it always begins with an inner need that leaves nothing to chance. When this need takes on substance, becoming autonomous, it takes off and reveals itself as transcendent. Within the interior, creative and spiritual life, things therefore begin with a need before turning into chance. They are first bound internally, before being unbound. If need is the condition of chance, then chance is the culmiation of need. What takes place in the creative process demonstrates this.

When an artist creates, they begin by working from the inside in order to take possession of their creativity. They are thus in possession of the need which will allow them to control their work. However, because of work, as they become increasingly interior due to increasing control, their work will take on its own autonomy and live by itself with its own inner need. It will then no longer be a need but a chance, surprising the artist themselves. This is what novelists describe.

When they create a novel with characters, if the novel really comes from within, the characters start living their own lives. If they are living their own lives, the novelist no longer directs them. Instead, they follow them. Having their own life, they become chance and need, chance because they are surprising, and need because they have their own logic. We are dealing here with the works of genius, the novelist managing to bring a world like that of Balzac's *La comédie humaine* (*The Human Comedy*) to life.

The question of chance and need has preoccupied philosophers for a long time. It can be found present in the works of Democritus during Antiquity, Diderot in the 18th century and Jacques Monod in the 20th century. Anxious to find a radical atheistic materialism, we find in all these thoughts the process of deducing the need for chance. When existence has nothing to hang on to, clinging to itself, it thus becomes active and responsible.

There is something magical about the idea that chance creates an inner need that the facts contradict. When beings are left to chance, forced to get by somehow, they do what they can. We are not dealing here with an inner need. This is mostly survival instinct fueled by the fear of dying. By making it very hard, this instinctive reaction does not promote a higher idea of existence or an intense desire to protect it.

You do not become an interior and creative being by living in constant disaster – with the proverbial knife to one's throat. On the contrary, one becomes an inner being by freeing oneself from disaster, that is to say, escaping the knife at one's throat. Every human being is at the crossroads of two forms of need. The first is external and arises from objective constraints. The second is internal and refers to an intimate need. When the objective need prevails over the inner need, by stifling the latter, human existence becomes harsh, constraining and sad. When the inner need becomes stronger than the outer objective need, human life knows freedom, creation and warm generosity. When the intimate life is stronger than the objective need, intimacy begins to come alive, many new unforeseen and unexpected connections begin to appear. Although these appear in the form of chances they are unexpected, they are not coincidence in the sense of an indiscriminate accident, but instead the result of a growing inner breath. The logic that governs the relationship between chance and need thus becomes clearer.

When you do not live from the inside out, chance and need clash. When they are opposed, they give birth to a harsh view of life based on the domination of chance by objective need. When one lives from within, freeing oneself more and more, the inner need opens up unexpectedly. Opening to the unexpected, the inner life makes one understand what finality is.

Philosophers and scientists have thought a lot about this without ever arriving at a satisfactory answer. Let us assume that reality pursues a goal. Everything being willed by a higher power pursuing a plan; reality becomes clear. It is possible to answer the question of why. However, as reality becomes the thing which a higher power directs according to a plan, freedom disappears. Let us eliminate all higher power with a plan when approaching reality. Let us just describe how things come together in view of this as they meet. Freedom is well respected, but the meaning disappears. This impasse is no accident.

When we look at things from the outside in, we can only end up with an image of reality in which things obey a need that is cut off from all chance, or a chance deprived of all need. Let us now start from the inside, in particular from the intimate need that we all feel lives within ourselves. The more this intimate need comes to life, the more it wants the unexpected in the form of the intimate life outside of it. Freedom does not contradict reason nor does freedom practice reason. Since reason wants freedom, reason and freedom grow together. The logic of life thus becomes clear. In existence, everything begins with bonding in a necessary way before unbonding. This link which binds and unbinds, the Ancients called this finality the *Logos*, Christianity the *Word*. We call it *Sense*. The idea is the same. The reality that comes from creation is there for creation. One has to interpret things in the light of the creation that binds and unbinds, and not by asking whether there is, or not, a purpose. We then see all the wealth that is contained within chance revealed.

9.3. When chance is not chance

When the ways in which creative chance presents itself to us begin with something strange, this is because there are chances that are in fact not chances.

What is strange is that these appear to us in the form of chance. Faced with this situation, there are several attitudes.

The first is that of Spinoza, who challenges this strangeness. A man walks down a street when all of a sudden a flowerpot falls on his head, killing him dead. Spinoza rejects the tempting idea that this was no accident; that the man died because he had to die. Let us stop thinking that a higher power caused this man to die or else we will end up in a superstitious delirium.

The second position is that of Cournot. When chance does happen, it is never completely by chance. For example, person A and person B meet at a dinner party. They were not expecting it. It is a happy coincidence, but it is not that surprising. Being part of the same social environment which hosts events often, there is a good chance that, at one point or another, the two will receive the same invitation. Hence, Cournot concludes that chance is an indeterminate event produced by a determined series which intersects. Persons A and B may themselves not have had any plans to meet, but the world to which they belong and which organizes events does, and as such there is no mystery to their eventual meeting.

The third position refers to structuralism. In the mechanical approach to things, every subject is conditional to their environment, which determines them according to an action-reaction principle, comparable to that of colliding billiard balls. In the living approach of things, it is different. When a subject is determined by an external element, it is not content to be dependent on it. It determines the other in return. As a result, a relationship is created between the subject and its environment. Becoming autonomous, it will determine both the environment and the subject. So in the end, the subject will not simply be determined by its environment, but rather, overdetermined through its relationship with itself.

In the 1960s, structuralism developed the concept of overdetermination in order to better understand the workings of society in a more perceptive way. Social life is often thought of as obeying an economic, political or cultural determinism. However, things are more complex in reality. It is the exchanges between living people that determine social life, not just economy, politics and culture. In practical terms, this changes everything. Everything is about relationships, about image and the games concerning image. If everything is a matter of image and "image games", it follows then that human relationships are a reflection of these games. Being the reflection of these games, they express their violence and creativity by going beyond the awareness of the subjects, as well as beyond their intentions, hence, the simultaneous appearance of determinism and chance. An important idea when delving deeper is that this becomes the basis of the *unconscious, initiation* and *incarnation*.

Psychoanalysis affirms that there is an *unconscious*. What we experience brings us back to what we are and vice versa. Events are not in front of us. We are the events. By recalling this, we realize that it is not possible to cheat ourselves as we might do with reality. One cannot pretend to be in love. When you do not love and pretend to love, sooner or later the truth finds a way out and becomes known, through setbacks, failures, missed opportunities and oversights. As if by chance, we fail, we forget, we miss that which we should notice. It is a little wonder, echoing the achemist tecahing *"Le même passe par le même"* (The same goes through the same). This is a law. The same that is on the outside must be the same on the inside. Otherwise, our *need* will cause conflicts so as to express what it does not like and connections to those which it does. On the surface, it will look like luck. In reality, these will be needs. Marx used this idea to explain that one cannot cheat social need. Freud used it in terms of desire.

If the notion of the unconscious helps one to better understand chance, then the *initiation* does so even more. We are not simply guided by our social lives and inner desires. We are called to rise to a higher spiritual life. When we refuse this life, this refusal comes back to us in the form of conflict. When we accept it, that acceptance comes back to us in the form of happy and inspired encounters. In any case, the relationship with a higher existence takes on the appearance of a coincidence that is unfortunate, hence the impression of chance.

According to Chinese wisdom, "The master comes when the disciple is ready". When the disciple is within wisdom, they have wisdom no longer in them, but before them. Jung calls this coincidence, between the inner and the outer, synchronicity. The ultimate work of initiation, incarnation, can then begin.

The notions of *incarnation* and embodiment are often confused, particularly when one thinks that in incarnation one must have a body. There is nothing material or bodily where incarnation is concerned. It is what happens when a figure comes alive; it is purely symbolic. Whether at an individual or collective level, the living principle that is at the heart of both the individual and the collective seeks to be incarnated. It comes to life when a correspondence is established between the outer and inner worlds. Proust understood this very well.

In human beings, there is that most incredible of faculties: memory. A mark of eternity and awareness within us, it ensures that we not only retain everything, but that we can recall everything, remember everything and thus make everything conscious at all times. We are not always aware of this creative potential within us. Our life being taken up by immediate utilitarian and egotistical concerns, we repress it. Only art, with its ability to look at the world for itself, can free our creative memory. This is always done through chance, either through visual or auditory resonance. As we are walking down the street, suddenly, three musical notes come out of a cafe or a car. As we like these musical notes, we stop and savor them. It does not take more than that to awaken our creative memory. Our inner need is so eager to live harmoniously that it seizes any outer harmony so as to resonate with it.

The churches that ring their bells know this principle. The bell is the instrument of resonance, and as such resonating a bell makes us resonant beings, helping us free our creative memory through the depth of auditory resonance as well as the depth of visual resonance.

Baudelaire called this "correspondence". We all have within us an interior geography referring to the countries and regions of our loves and our emotions. When love and emotion take shape in the exterior world, through the tenderness of things, hearts and ambience, it takes little extra to awaken this interior geography and let it express itself. Whether it is Proust or Baudelaire, in both cases, the principle remains the same.

The world being the world of life where everything is lived, there exists a basic correspondence between Life with a capitalized "L" and life with a lowercase "l". Everything is given for this correspondence to be established, because in principle, reality is thought of as a great correspondence, a great resonance, a great mirror, a great echo, a great radiance and a great wave propagation. Everything is given except for one thing: the little nudge which will cause it all, suddenly, for unknown reasons, the outside and the inside will meet, by making Life with a big "L" communicate with life with a little "l", the microcosm of the heart with the macrocosm of infinity. Art is the story of encounters of this nature.

Nature and the poet are destined to meet. They are called to one another because Nature awaits *her* poet who, for *his* part, is likewise in search of Nature. The same goes for history. The greatness of history awaits a great person who, in turn, seeks out historical greatness. The encounter takes place for mysterious reasons through details which are not details and through chances which are not by chance. In depth, the universal link which allows everything to meet, to correspond, to radiate and to resonate already exists, but we do not know how to see it.

Reality is made up of three elements. If there is the world, if there is *us* in front of the world capable of representing the world to ourselves, then there is the consciousness of the world toward us, and the consciousness of this consciousness. This is infinite consciousness. It is this that binds everything and allows everything to be linked. Plato called this the Enlightening, Karl Jaspers the Encompassing. This is the first of any gifts, allowing the information of the universe and that of humanity to meet. Almost as if one overhears, by chance, a benevolent host organizing a party for the universe and humankind to meet. This is what Basarb Nicolescu calls the Hidden Third, which allows for everything to flow and meet.

9.4. When chance comes from elsewhere

Chance belongs to this world. However, it is not just of this world. Sometimes it carries things from elsewhere along with it. When you embark upon the inner adventure, this is always the case. Let us call the "divine" the mysterious fact of an elsewhere which speaks to the depths of our being, as the being of life, when we enter into an interior spiritual path, an elsewhere which is both infinite and intimate begins to manifest itself through *signs*, *happenchances* and *graces*.

Reason does not like *signs*, and it is not completely wrong to harbor this mistrust. The sign after all may be a pathological projection. The superstitious soul does not stop looking for signs. Wanting to have confirmation from the outside of what one feels inside oneself, it never ceases to look into reality for elements that comfort it. When it finds them, these elements become signs. They refer to what it has in itself. This way of submitting the outer world onto oneself, and what one feels, is problematic. Those who practice this firmly believe that their signs come from heaven, even though they were invented by themselves. Therein lies the problem. It is not just an illusion. It is a hallucination and, behind it, a delirium. However, not all signs are of this sort. There are some that relate to something other out of a need for reassurance.

It is fine to refer to something else. This is an opening. From this opening, it is the elsewhere that speaks, and behind the elsewhere, the privacy of ourselves. When we begin to live with elsewhere, the world and life are populated with signs. It is not a coincidence. The world and life begin to become a master teaching us in a thousand ways that we must go further. The sign is therefore not misleading. It is an awakening. It does not make sense to reassure us by bringing us home. Rather, it has a disorienting effect by transporting us beyond ourselves.

The sign is at the heart of life's most profound experiences. When we awaken to the inner life, we receive signs in the form of appeals that are both urgent and discreet. These callings invite us to change. Coming from the new, they invite us to be like it, not only free, but seriously free. Therefore, they are linked to an inner need, since they are seriously free and, at the same time, open to the unexpected.

Alongside the signs, these divine chances, there are the happenstances, other divine chances. Vladimir Jankélévitch speaks on this subject of "fairy happenings". In Aristotle, happenchance is called *kairos*. One is tempted to translate this term as "opportunity", but that only faintly captures what is taking place within *kairos*, a truly happy coincidence wherein everything falls into place just right; we are no longer following the ordinary course of things within *kairos*, but instead the extraordinary.

For Aristotle, chance, by definition, is the reflected image of a degraded world. While the substance has cause within it, the accident has its own cause outside of it. Having no causality within itself, chance is not the master on board. Depending on what is causing it, it is subject to whimsy, hence its chancy existence. Unlike an accident, which has its own cause, the substance is anything but chancy. Still, it is no less paradoxical. When its name is Life with a capital "L", giving life to everything without anything giving it life, without a cause – this is chance. It is not, however, by chance. The reason is dialectical. For the cause to be a cause, it needs for there to be a cause without a cause. Are there infinite causes at the origin of the cause? This one can then disappear. Chance, which is existential misery *par excellence*, is also the divine *par excellence*. If there is a small chance, then it follows that there is a big chance. When a chance is great, it is not only great but magnanimous in its capacity for giving. It is possible to experience it. An opportunity always presents itself to us as a chance. Nevertheless, being generous, this chance is no longer a chance but an opening, a gift. In fact, it is a chance from elsewhere. With *kairos*, Aristotle makes this understood. If there are insignificant chances, there are significant chances: these are those which are not locked into chance. Free, they speak of the being that Aristotle also calls God. On the earthly plane, they talk about happenstance. If free and divine chance did not exist, chance would be encompassed by chance, and everything would be necessary. Everything being necessary, nothing really would be, and in fact everything would be chance. The profundity of free and divine chance preserves the world by allowing for there to be a balance.

At this juncture, it is appropriate to speak of Grace. Coming from elsewhere, this is not chance, and yet, it is connected. Completely unexpectedly, sometimes people take pity and instead of forgiving nothing, they give thanks. Also totally unexpectedly, sometimes people are seized with genius and instead of being dull, they start to be elegant, even supremely elegant. Finally, totally unexpectedly once again, it sometimes happens that where there is darkness, light manifests, and where there is light, the light begins to shine superabundantly. The human heart and human life are capable of totally unexpected actions that humans and normal life do not explain. To be honest, there are possibilities in the human heart as in life that go against the course of things, by totally reversing them. Without a gift from elsewhere, this reversal is not possible, or even understandable.

Grace is not of the order of chance; chance is not that which produces Grace. However, it tells us something, by chance, when it tells us about what (eluding all logic) is not a need but a gift from elsewhere. The world, like human life, is not simply governed by material or even human logic. Without simply being governed by material or human logic, they are not left to their own fate in a horizon without light either. This is shown by Grace; this is an unexpected gift from elsewhere and appearing outside all normal logic without being absurd. There is more to existence than we might think. By studying chance, we begin to become aware of it.

At the beginning, we always start by being afraid of and rejecting chance, seeing in it a troublemaker that vandalizes logic and thought. However, there always comes a time when, intrigued by it, we change our minds by starting to think that it is just too illogical to simply be illogical. Realizing then that there are chances that are not, we end up discovering what it is. Chance is explained by consciousness and inner life and not through linear logic. We do not believe that reality is spiritual and that it relates to some other source of existence that manifests itself through all kinds of unexpected and creative events. Yet, this is what chance invites us to think.

10

Chance and the Creative Process

10.1. Introduction

I would like to start this chapter by mentioning a few words about my current position as it directly relates to the content of this chapter. I present myself as an Artist Researcher[1,2] (Magrin-Chagnolleau 2014), that is to say, I pursue creative artistic activities while at the same time research activities, at CNRS, based around this artistic activity. I work within the paradigm of Creative Research (Magrin-Chagnolleau 2014)[3], which means that my creative artistic work and my research work are intimately linked. My research is related to my creative artistic work, and more generally, it is related to Art, Esthetics and Philosophy of Art. While my creative artwork is also linked to my research, it represents the grounds for my research and also constitutes a research by itself, whether it is of an esthetic nature or a more general philosophical nature. Sometimes these creations allow me to express some of the results of my research in a way that is more relevant than an article.

On January 1, 2019, I joined the PRISM laboratory in Marseille, a new UMR[4] (7061) or *joint research unit*, created to precisely explore the links between Art and Science, creative research being one of the ways by which to explore this link. Within the framework of this creative research paradigm, I am particularly interested in the phenomenology[5] (Husserl 1992) inherent to the creative process, that is, the

---

Chapter written by Ivan MAGRIN-CHAGNOLLEAU.

1 *L'Artiste Chercheur*, first issue of the journal *p-e-r-f-o-r-m-a-n-c-e.org*, vol. 1, 2014.

2 Magrin-Chagnolleau, I. (2014). Éloge de l'artiste chercheur. *p-e-r-f-o-r-m-a-n-c-e.org*, 1.

3 Also called Art Research.

4 Unité Mixte de Recherche (Joint Research Unit).

5 Husserl, E. (1992). *L'idée de la phénoménologie*. PUF.

creative process as a lived experience. How can we communicate the creative process as experienced and perceived by the person who is the actor of this process?

The question of the intervention of chance in this creative process is therefore an essential question, and it is this question that I seek to address in this chapter. I will begin by discussing the term "chance" and a number of related concepts. Next I will look at the term "creation" and its constellation of related notions. I will then reflect on the intervention of chance in the artistic creative process. Following this, I will address the question of an artwork made in the present moment. I will wrap up this chapter with a conclusion.

10.2. Chance

I have gotten used to starting an investigation with a semantic exploration of a word as taken from those definitions offered by *Le Grand Robert*[6] (Rey 2017), since it is through a dictionary that the history of the word can be seen, and as such the different concepts attached to it, its various occurrences in the literature, wherein each occurrence sheds a different light on the word. In the case of the word chance, we find that its meanings and occurrences are numerous.

First, the etymology of the French word for chance – *hasard* – (the English word "hazard" can also be used in the same context) refers us to an Arabic word, which has come down to us via Spanish. The origin of the Arabic word is controversial and could mean either "game of dice", "flower" or "orange blossom" because ancient dice once had a flower marked on one of the faces.

The oldest definition (13th century) tells us: "A dice game practiced in the Middle Ages". And a definition from 1538 describes: "Game of chance: game where the winner is chosen at the end of a series of 'moves' randomly producing results, and where calculation, skill ... plays no part".

So, from the beginning, the notion of chance is linked to games. There are games of chance, where everyone has the same chances, and where the skill of each plays no significant role, and there are the other games. We also immediately see a link established between chance and **random**, and by extension **randomness** (the randomness of life).

In the 15th century, we have the following definition: "Case, fortuitous event; unexpected and inexplicable combination of circumstances". So now we have the intrusion of notions such as **fortuitousness**, the **unexpected** and the **inexplicable**.

---

6 Rey, A. (2017). *Le Grand Robert de la Langue Française*. Le Robert.

In the middle of the 16th century, we have: "Fictitious cause of what happens without apparent or explicable reason". Here, again, we find the notion of what is explainable and what is not, and is therefore **inexplicable**. There is also an opposition between that which has a reason for being and that **which has no** *apparent* **reason for being**.

In philosophy and science, we have at the same time: "Character of what happens outside objective or subjective norms, that which falls under the laws of probability and is not deliberate". Here, we find the first mention of the notion of the 'norm', and how chance exists outside of the norm, and is therefore **anormal**. We also notice the link between chance and **probability**. And finally the notion of deliberateness, chance as such is therefore **non-deliberate**.

Montaigne wrote in 1580: "Without a determined direction; without deciding the place, the situation". Here, it is rather the idea of **non-decision** that is put forward. We could also say **non-intention**.

And Fénelon states in 1695: "Without foreseen nor foreseeable evolution; without specific modality". Here, it is the evolution that is **unpredictable**, or the modality that is **undefined**.

We also have the adverbial phrase "by chance" that can be translated as "**accidentally, fortuitously**". Here, the notion of **accident** appears.

I would like to end this exploration of *Le Grand Robert* and its definitions of the word "chance", with the help of a few quotes that will shed some light on the remarks made in this chapter. I will start with a maxim from La Rochefoucauld[7]:

> Although men flatter themselves with their great actions, they are not
> so often the result of a great design as of chance. (La Rochefoucauld
> 1999, 57)

Here, we have an opposition of the word "chance" and the word "design", that is to say, something planned. Chance is therefore **unplanned**.

There are two other quotes concerning the word chance that I find very interesting. The first is by Romain Rolland in *Jean-Christophe*[8]:

> [...] chance always knows how to find those who know how to use it.
> (Rolland 2007)

---

7 La Rochefoucauld, F. (1999). *Maximes*. Garnier Flammarion.
8 Rolland, R. (2007). *Jean-Christophe*. Albin Michel.

The second is from Albert Camus in *La Peste*[9] (The Plague):

> His only task, in truth, was to give opportunities to this chance which, too often, is disturbed only when provoked. (Camus 1972)

These two quotes establish a link between chance and the ability of each and every person to use it and provoke it. Therefore, **chance needs to be provoked**.

Voltaire, in his *Philosophical Dictionary*[10], brings us back to the fact that that which we call chance is **that which we do not understand**, or that which we do not understand *yet*:

> What we call chance can be no other than the unknown cause of a known effect. (Voltaire 1994)

I would like to continue on this semantic and philosophical exploration with two more quotes about the word "chance", which emphasize the question of point of view, that is to say, that **there can be no chance without a subject who is concerned by the consequences of this chance**.

The first is by Henri Bergson in *Les deux sources de la morale et de la religion* (The Two Sources of Morality and Religion)[11]:

> A huge tile, blown off by the wind, falls and knocks down a passerby. We say it's a fluke. But would we say that if the tile had just broken on the floor? Perhaps it is because we vaguely think, what if a person had been there, or because for one reason or another, this special point on the sidewalk is of particular interest to us, such that the tile seems to have chosen this spot to fall. In both cases, there is only chance because there is human interest involved and because events took place in such a manner that it seems the man had been taken into consideration, either with a view to rendering him a service, or the converse, with the intention of harming him (...) Chance is therefore a mechanism, behaving as if it had an intention. (Bergson 2012)

The second is by Paul Valéry in *Variété* (Variety)[12]:

> "Everything else – everything that we cannot assign to the thinking man or to this Generating Power (nature) – we offer to 'chance' –

9 Camus, A. (1972). *La Peste*. Folio.

10 Voltaire. (1994). *Dictionnaire philosophique*. Folio.

11 Bergson, H. (2012). *Les deux sources de la morale et de la religion*. Garnier Flammarion.

12 Valéry, P. (2002). *Variété III, IV et V*. Folio Essais.

which is itself an excellent addition to our vocabulary. It is expedient to have a name which allows one to express a remarkable thing (by itself or by its immediate effect) has taken place just as easily as it might not have done. However, to say that a thing is remarkable is to introduce a man, a person who is particularly sensitive to this trait, and they are the one who will attribute all the remarkable things to it. What does it matter if I don't have a lottery ticket, with the exact same numbers as that called out in the draw? (...) There is no chance for me in the draw (...). (Valéry 2002)

Finally, to conclude this exploration of the word "chance" with the help of *Le Grand Robert*, I cannot help but end with a quote from Jacques Monod in *Le Hasard et la Nécessité* (Chance and Necessity)[13]:

We say that these alterations (of the "genetic text") are accidental, that they happen at random. And since they constitute the only possible source of modification for the genetic text, in turn the sole depositary of the hereditary structures for the organism, it necessarily follows that chance alone is at the source of all novelty, of all creation within the biosphere. (Monod 2014)

This quote, even if its content has been questioned somewhat by recent advances in epigenetics, is interesting because it links the word "chance" with the word "accidental", and hypothesizes that all creation (in the biosphere) and all novelty (in the genetic text) is necessarily the result of chance.

10.3. Creation

Let us now begin a semantic exploration of the word "creation". Here again, *Le Grand Robert* provides several definitions. The first definition is more theological: "The act of giving existence, of drawing from nothing". Here, "creation" is directly linked to an action. It is through **action** that we create. By creating, we give **existence**. Before being created, the created thing does not exist.

There is also this notion of "**drawing from nothing**". What is nothingness in this definition? If we bring the two elements of this definition together, we could say that nothingness is that which is non-existent. Or else nothingness is the field in which all is possible, but which has not yet been acted upon, that is to say, nothing has ultimately been chosen.

---

13 Monod, J. (2014). *Le hasard et la nécessité*. Points Essais.

To further illustrate this theological definition, *Le Grand Robert* gives us several interesting quotes. The first comes from Montesquieu in *De L'Esprit des Lois* (The Spirit of the Laws)[14]:

> If creation appears to be an arbitrary act... [it] presupposes rules that are as invariable as the fatality of the atheists. (Montesquieu, 1993, 1)

Montesquieu speaks to us here on the **rules of creation**. Creation therefore has rules, and what is more: invariable rules. He uses here the contrasting term "arbitrary", which implies having no rules. Is chance therefore arbitrary? He also relates these rules and their invariability to what he calls the "fatality of atheists". Here, we find ourselves at the center of the question of free will versus predestination. When there is creation, does this come from free will, from a completely free choice on the part of the creator, or is everything predetermined and following invariable rules?

Jean-Jacques Rousseau (1959–1995), in a letter to M. de Beaumont[15], reminds us how difficult it is to understand and apprehend **this journey from nothingness into something**:

> Now, the idea of creation, *viz.* the idea by which we conceive that, from a simple act of volition, nothing becomes something, is, of all ideas that are not evidently contradictory, the least comprehensible to the human mind. (Jean-Jacques Rousseau, letter to M. de Beaumont, November 18, 1762)

Finally, Paul Claudel (1968–1969), in his *Journal*[16], enlightens us on the difference between **creating from nothing** and **creating in the continuation of what has already been created before**:

> There is a difference between the 1st day of Creation and the others. On this first day God creates from nothing, out of nothing. On the other days God creates in the middle of something. His creative activity is in a way conditioned by the other things He has already created. It is a matter on which He acts, that He provokes. (Claudel, Journal, September 13, 1922)

This question is essential in relation to the questions of the link between chance and creation by differentiating a creation that is primary from a creation that would

---

14 Montesquieu. (1993). *De L'Esprit des Lois*. Garnier Flammarion.
15 Rousseau, J.-J. (1959–1995). *Œuvres complètes*, Volumes I to V. La Pléiade Gallimard.
16 Claudel, P. (1968–1969). *Journal*, Volumes I & II. La Pléiade Gallimard.

continue a work already begun. This question arises in both artistic creation and in research.

Next we come to a more common definition that makes reference to the activities of man rather than that of any divinity: "**Action** to do, to organize (something that did not yet exist)".

Here again, a few quotations help shed new meaning of the word "creation". Let us start with a quote from Eugène Delacroix (1996) as taken from his *Journal*[17]:

> It is generally acknowledged that what is known as creation in the great painters is only a special manner in which each of them saw, coordinated, and rendered nature. (Eugène Delacroix, *Journal*, March 1, 1859)

Any creation is thus a way of **sharing one's own vision of the world**. For there is in this quote from Delacroix, not only the action of seeing but also that of coordinating, that is to say, a way of **giving meaning**.

Edmond Jaloux, for his part, speaks to us in *La Chute d'Icare* (The Fall of Icarus)[18], on the link between creation and action but also on **continuous creation**:

> To act is a continuous creation. Nature is constantly creating forms that have no value unto themselves, but it is the sum total of these infinitesimal creations that is life. (Edmond Jaloux 1935, p. 217)

At the heart of the question posed here is the creation of nature, but also the idea of **incremental creation**, that is to say, in art, as in research, it is by small infinitesimal contributions that a work, a new theory, etc., comes into being.

The next quote I would like to share is from Max Jacob in *Conseils à un jeune poète* (Advice to a Young Poet)[19]:

> That which saves art is invention. Creation only exists where there is invention. Each art has its inventions. (Max Jacob 1945)

In this quote, Max Jacob links creation to **invention**. There is creation when something new is invented, which of course raises the question of novelty. How can we define that which is new, for example, when it takes place within incremental creation? Or an equivalent in research, when there is no paradigm shift, but only an

---

17 Delacroix, E. (1996). *Journal, 1822-1863*. Plon.

18 Jaloux, Edmond. (1935). La Chute d'Icare. In *La Revue des Deux Mondes*.

19 Jacob, Max. (1945). *Conseils à un jeune poète*. NRF.

infinitesimal contribution, compared to state-of-the-art contributions? And conversely how can we assess the interest, the relevance, the aesthetics of something that truly represents a paradigm shift? Can we compare a paradigm shift in art or research to a genetic mutation?

Albert Camus, for his part, gives us, in *Le Mythe de Sisyphe* (The Myth of Sisyphus)[20], a more existential thought:

> Of all the schools of patience and lucidity, creation is the most effective. It is also the staggering evidence of man's sole dignity: the dogged revolt against his condition, perseverance in an effort considered sterile. (Camus, 1885)

Therefore, creation, for mankind, would be a way of fighting against the inevitability of death, a way of leaving a **trace**, a **testimony** of the lived experience.

I would like to end with a quote from André Malraux in *Les chênes qu'on abat...* (Felled Oaks; Conversation with De Gaulle)[21]:

> Creation has always interested me more than perfection. (Malraux, 1971)

This quote reminds us that it is perhaps far more important to *try* rather than *want* to be perfect, whether in artistic creation or in research. The most important thing is to create something, not to strive for perfection.

10.4. Chance in the artistic creative process

After attempting to somewhat clarify the terms "chance" and "creation", I would now like to address the connection between the two. What link is there between chance and the creative process? How does chance intervene in creation?

In this section, I will be mainly interested in artistic creation, and as such I will mainly rely on my own experiences, since to talk on this intervention of chance in creation, I can see no better way to approach this question other than from a phenomenological point of view[22] (Husserl 1992), that is, from the point of view of the lived experience. Indeed, only a lived experience can begin to clarify what can be considered as chance in a creative process. As we have seen with Henri Bergson

---

20 Camus, A. (1885). *Le Mythe de Sisyphe*. Folio Essais.
21 Malraux, A. (1971). *Les chênes qu'on abat...* . NRF.
22 Husserl, E. (1992). *L'idée de la phénoménologie*. PUF.

and Paul Valéry, chance is a very subjective notion, dependent on a point of view, and therefore on the experiences that have led to this point of view.

It would be interesting, to take this reflection further, to consider an experimental protocol that aims, for example, with the help of elicitation interviews[23] (Vermersch 2017) or phenomenological interviews[24] (Bitbol and Petitmengin 2017), to question other artists, and also researchers, on their lived experiences of the creative process in which chance has been perceived as a constitutive or even a defining element. We could then speak of the phenomenology of chance, or to put it in the form of a question: How is chance perceived in any form of creation from the point of view of the lived experience?

For the moment, let us come back to my own lived experience, and my own relationship with this notion of chance as manifested in my creative works. To this end, I would like to draw on several examples, each illustrating in a different way what the word "chance" can *re-comprise* within the creative process.

The first is an example from photography, and concerns a series of photos called *Muir Woods Spirits*, which is regularly exhibited in galleries and which has been the subject of a book publication[25] (Magrin-Chagnolleau 2017).

A friend and I visited Muir Woods National Monument Park a little way north of San Francisco. I had with me a camera as I so often do. Once in this extremely majestic location, I started taking some very classic shots, playing with the perspective effects of these gigantic trees.

At one point, I realized that this classic type of photo did not suit this place, which gave off something quite different. I found myself relating to it more as a natural cathedral. I then started taking pictures while moving my camera or its zoom or both simultaneously. It was then, from that moment on, only a matter of experimenting with movement, shutter speed and aperture.

What interests me here is the precise moment at which I started to experiment with these manipulations. It was a process which at the time was not a part of my usual photography process. I had undoubtedly seen it in other photographers, and maybe even used it once or twice without really paying attention, for example, when I take pictures "by chance", without looking through the viewfinder. But in this case, I cannot remember the moment when I would have decided: "Okay, now I'm going

23 Vermersch, P. (2017). *L'Entretien d'explication*. ESF Éditeur.

24 Bitbol, M. and Petitmengin, C. (2017). Neurophenomenology and the elicitation interview. In *The Blackwell Companion to Consciousness*. 2nd ed. John Wiley & Sons.

25 Magrin-Chagnolleau, I. (2017). *Muir Woods Spirits*. Aloha Edition.

to take pictures while moving the camera". It may have been that the first photo with camera shake was taken by accident, meaning that I moved unintentionally, and was then captivated by the result. Or that this decision to move the camera while taking the photo arose out of my subconscious, in that present moment, an intuition, etc. In any case, there is something involved here that was not premeditated, which had not been decided, and something that had not been foreseen, which was accidental. If we refer to the first section of this chapter, we can see that we are precisely in the semantic field of the word "chance".

The second example also comes from photography, and concerns a more recent series of photos called *Handscapes*, which is just starting to be shown in galleries and whose book release is imminent[26] (Magrin-Chagnolleau 2020).

I went to the heart of Auvergne one Summer for a training course. To travel there I took a small regional train, which runs through beautiful forests and canyons. Having become fascinated by the landscapes that I crisscrossed, I resolved to take photos of these sublime landscapes on my return. So on the way back, I took out my cell phone and took my first photo. I observed that in this photo, in addition to the photographed landscape, I could also see the reflection of my hands and my phone in the train window.

At first I tried to find a way of avoiding this reflection, but quickly realized that there is something fascinating about seeing a photo and at the same time the tool that made it possible to take said photo, as well as the hands holding said tool. I then made the decision to continue taking pictures with apparent reflection.

Again, what particularly interests me here is that when I took the first photo, I did not plan to photograph my reflection. I was just taking a photo with the intention of capturing a beautiful landscape the way I saw it. Herein lies the fruit of chance that composed for me a photo where we see at the same time a landscape, a cell phone and my hands. Of course, it is no coincidence that this reflection appears. It is the result of a very specific physical phenomenon. But the chance comes about through the fact that I had not anticipated for this physical phenomenon. I had not foreseen it. I had not planned for it. Nor did I choose it. It occurred without my knowledge. From the point of view of my lived experience, this is "chance". This brings us back to those quotes from Henri Bergson and Paul Valéry, as in the previous example, who state there can be no chance without a subject.

I would like to take a third example that is perhaps even more surprising. In the late 1990s, I was in New York working as a researcher at Bell Labs, which later

---

26 Magrin-Chagnolleau, I. (2020). *Handscapes*. Aloha Edition.

became AT&T Labs. I was living in New York then, and I was taking beginner watercolor classes at Cooper Union, then one of New York's top art schools. I made six beginner watercolors, which were not particularly fantastic, but which initiated a practice, and above all a very special relationship to this art medium that I still cherish. The watercolors have since been stored in a cardboard folder with a few blank sheets. Upon my return from the United States, this box was stored in an attic in a family country home.

About 15 years later, while tidying up the attic I came across this box. I opened it and found that it had absorbed a little moisture, and that most of my beginner watercolors had been damaged. At the same time I also discovered a very surprising thing. Thanks to the humidity, through capillarity and sedimentation of the pigments, an absolutely magnificent watercolor had "composed itself" on one of the blank sheets, a kind of meditative landscape[27] (Magrin-Chagnolleau 1998–2018). Again, we can explain very precisely what happened from a physics point of view. But who could have planned for such a thing, without ever having experienced it, "by chance", first hand? From the point of view of my lived experience, it was truly chance that made this watercolor possible.

I could cite many other examples of the same type, from my artistic practices and also some from my experience as a researcher, but it seems to me that these three examples already allow us to fully understand the subject of this chapter.

Indeed, these examples illustrate the close links between chance and creation, and also underline, like Henri Bergson and Paul Valéry, the phenomenological part to this relation, that is to say, the fact that there is always a subjectivity to interpret a lived experience as the intervention of chance.

10.5. An art of the present moment

I would like to end this chapter by talking about an experiment that I have been conducting for many years around an art of the present moment, that is to say an art practice that is shaped in the moment, without premeditation, without predictability, an artwork where chance plays an important role.

This experience has already taken several forms. It is currently being developed around two flagship projects.

The first is called "Around the Lake". From 2012 to 2017, I lived near a lake. I very quickly got used to going for walks there, and very quickly I got used to these

---

27 Magrin-Chagnolleau, I. (1998–2018). *Sedimented Landscape #1*. Watercolor.

walks taking me around this lake. I quickly made the decision that I was going to take these walks around the lake so as to explore the creative process. It was done in stages. First I took notes, on my first few walks, of any thoughts that came to me as I walked around the lake. Then some of these ideas led to other forms of experimentation. I started taking pictures during these walks. I then captured audio recordings of the sounds I would hear. I then took to taking short videos composed of several moments during these walks. It was then that I realized the impact of a regular walk on creativity. It was also around this time that I joined the Walking Artists Network (WAN)[28], a network of research artists who explore this link between walking and creativity, and who make walking one of the practices of their creative and research process.

At the end of those 5 years I had amassed a considerable amount of material, a sort of dough that I now had to knead into shape. The material is itself a kind of work already, but it is also the formatting that gives meaning to this work. So I decided to produce a number of "objects" from this material. First, compiling a book that was written during these hikes as well as some photographs. Then another book essentially comprised photographs. These two books are in the process of being produced and will hopefully be published at some point in 2021. I would also like to create a website where I can collate all the texts, photos, sound and video files into one place. Finally, I would like to create an immersive exhibition where these different materials would be arranged as a coherent whole, and where the viewer of the exhibition wanders through it all, a bit in the same manner I wandered around this lake myself.

What is fascinating about this kind of experience, beyond all the deep intuitions that arise during regular walks, is the preponderant role that chance plays in the moment of creating artistic objects. Even if, as we have already seen, there must be a subject in order for there to be chance, and even if we can infer that the thoughts that arise, the desire for photos, sound recordings or video files, are triggered by volition, it is what happens in the moments following this triggering that belong to chance.

For example, if I decide, without always knowing why, that I am going to start recording sounds, I do not know what will happen between the triggering of the recorder and the moment I stop it. While it may have been a bird's song that was the trigger, I have no idea what will happen next. I may see children playing, or a woman walking her dog, or ducks that will fly away as I approach them.

It is the same with video. Although I might have a good reason for starting to film, often at times the reason is simply that I think about it then. Nevertheless, even

---

28 https://www.walkingartistsnetwork.org.

though I am filming where I am, often the path right in front of me, what happens after the start of the recording is spontaneous. I may take a detour to avoid an unusual obstacle, or else be attracted by something that takes me off my normal route.

The second project is called "The Haiku Project". I have been writing haiku for a long time now and with it the idea to create an entire project around the notion of haiku. I wanted to explore forms of haiku other than poetry, including photographic haikus, video haikus, musical and sound haikus. For this purpose, I opted to ease the constraints of haiku so as to retain two aspects: the simplistic form and the idea of an artwork in the present moment, not premeditated, but that which is made in the here and now.

Here, again the question arose of objects made from all the material accumulated over time. As with the previous project, I chose to start with that which seemed easiest to do, a book of poetic haikus that also contains some photographic haikus, as well as a book that mainly contains photographic haikus. These two books are also in production and will likely be published in 2021. I also intend to create a website to host a lot of the material collected as part of this project. Again, I would also like to stage an immersive exhibition that not only allows the viewer to wander through all this material but to also experience the haiku process in different forms, this creative process that is within the moment.

Here, again chance plays a considerable role in the creative process. There is of course always the question of the trigger. What makes me choose this time to make a haiku, and what makes me choose this form of haiku over another? This is why this artistic practice is extremely phenomenological, since one can assume that it is something that happens at the level of the lived experience that triggers this impulse to do a haiku as well as to choose in what format.

For some time now, I have been considering taking this notion of chance even further, by setting up a small system that would randomly choose the moments of the haikus as well as their medium. It would then be interesting to compare the two forms of practice. I am not sure, however, that it is all that interesting to disconnect the practice of haiku from the phenomenological experience that triggers it. Although in this second case, there would still be a lived experience, but it would be of quite a different nature.

10.6. Conclusion

I have therefore proposed in this chapter a semantic exploration of the word "chance" and the word "creation". Next I have tried to show, using examples from my own artistic practice, the part played by chance within the artistic creative process, and the extent to which chance was or was not a constituent element of this process. I then presented two artistic experiences that explored an artwork of the present moment, that is to say an artwork in which, by definition, chance plays a significant role.

While concluding this chapter, I have formed a new idea, mentioned above, to take this exploration of the link between chance and creation further. For some time now I have been carrying out experiments around the phenomenology[29] (Husserl 1992) of the creative process[30] (Magrin-Chagnolleau et al. 2019). I do this through elicitation interviews[31] (Vermersch 2017) and phenomenological interviews[32] (Bitbol and Petitmengin 2017) which I conduct with several artists based around their creative process. During these interviews, I suggest that artists choose a moment in their creative process, in connection with a recent creation, which they find particularly interesting. On the other hand, I could also ask them to choose a moment when it seems that chance played an important role. This would allow me to investigate and characterize the various notions different artists attach to the word "chance". In any case, it would certainly make for interesting further explorations.

10.7. References

L'Artiste Chercheur (2014). *p-e-r-f-o-r-m-a-n-c-e.org*, 1.

Bergson, H. (2012). *Les deux sources de la morale et de la religion*. Garnier Flammarion, Paris.

Bitbol, M. and Petitmengin, C. (2017). Neurophenomenology and the elicitation interview. In *The Blackwell Companion to Consciousness*, 2nd edition. John Wiley & Sons, Oxford.

Camus, A. (1885). *Le Mythe de Sisyphe*. Folio Essais, Paris.

Camus, A. (1972). *La Peste*. Folio, Paris.

Claudel, P. (1968–1969). *Journal*, Volumes I & II. La Pléiade Gallimard, Paris.

Delacroix, E. (1996). *Journal, 1822–1863*. Plon, Paris.

Husserl, E. (1992). *L'idée de la phénoménologie*. PUF, Paris.

---

29 Husserl, E. (1992). *L'idée de la phénoménologie*. PUF.
30 Magrin-Chagnolleau, I., Nahoum, R., Weisen, M. (2019). Le projet sipapu: vers une micro-phénoménologie de la création. In *Les enjeux cognitifs de l'artefact esthétique, tome 2*, Lambert, X. (ed.). L'Harmattan.
31 Vermersch, P. (2017). *L'Entretien d'explicitation*. ESF Éditeur.
32 Bitbol, M. and Petitmengin, C. (2017). Neurophenomenology and the elicitation interview. In *The Blackwell Companion to Consciousness*, 2nd edition. John Wiley & Sons.

Jacob, M. (1945). *Conseils à un jeune poète*. NRF, Paris.

Jaloux, E. (1935). La chute d'Icare. In *La revue des deux mondes,* November.

La Rochefoucauld, F. (1999). *Maximes*. Garnier Flammarion, Paris.

Magrin-Chagnolleau, I. (1998–2018). *Sedimented Landscape #1*. Watercolor.

Magrin-Chagnolleau, I. (2014). Éloge de l'artiste chercheur. *p-e-r-f-o-r-m-a-n-c-e.org*, 1.

Magrin-Chagnolleau, I. (2017). *Muir Woods Spirits*. Aloha Edition, Paris.

Magrin-Chagnolleau, I. (2020). *Mains paysages*. Aloha Edition, Marseille.

Magrin-Chagnolleau, I., Nahoum, R., Weisen, M. (2019). Le projet sipapu : vers une micro-phénoménologie de la création. In *Les enjeux cognitifs de l'artefact esthétique, tome 2*, Lambert, X. (ed.). L'Harmattan.

Malraux, A. (1971). *Les chênes qu'on abat...* NRF, Paris.

Monod, J. (2014). *Le hasard et la nécessité*. Points Essais, Paris.

Montesquieu. (1993). *De l'esprit des lois*. Garnier Flammarion, Paris.

Rey, A. (2017). *Le Grand Robert de la langue française*. Le Robert, Paris.

Rolland, R. (2007). *Jean-Christophe*. Albin Michel, Paris.

Rousseau, J.-J. (1959–1995). *Œuvres complètes*, Volumes I–V. La Pléiade Gallimard, Paris.

Valéry, P. (2002). *Variété III, IV et V*. Folio Essais, Paris.

Vermersch, P. (2017). *L'Entretien d'explicitation*. ESF Éditeur, Paris.

Voltaire. (1994). *Dictionnaire philosophique*. Folio, Paris.

Walking Artists Network (2021). Home Page [Online]. Available at: https://www.walkingar tistsnetwork.org.

PART 2

Randomness, Biology and Evolution

Epigenetics, DNA and Chromatin Dynamics: Where is the Chance and Where is the Necessity?

11.1. Introduction

There are many ways to tackle challenging scientific questions. Some researchers build elaborate plans to solve a problem, whereas others leave a little more room to chance. The word "*serendipity*", coined by Horace Walpole in the 18th century, describes inventions and discoveries that result from chance (Walpole 1754). The history of science and discovery is full of memorable examples, from Christopher Columbus's discovery of the American continent to Alexander Fleming's accidental stumbling on the first antibiotic, Penicillin. Fleming's untidiness is often cited as the cause of the chance discovery and he himself admitted that he had not planned to revolutionize medicine by discovering the world's first bacteria killer (Fleming 1929). Indeed, scientific discovery is a relatively haphazard and sometime unintentional process, but could it be that biological processes themselves also rely on chance and necessity?

11.2. Random combinations

If we zoom in to the microscopic level, the study of biology's building blocks can be extremely enlightening. The molecules of deoxyribonucleic acid (DNA), in the nucleus of each of our cells, carry the heritable genetic information that instructs the unique features of a living organism (Hershey and Chase 1952). The sum of the entire genetic material of an organism is referred to as the "*genome*", but we spent

Chapter written by David Sitbon and Jonathan B. Weitzman.

much of the 20th century focusing our attention on the constitutive components, which we refer to as "*genes*" (Pearson 2006). The gene itself has undergone a series of re-definitions, from a conceptual unit of inheritance to a physical entity along the genome. Today, biologists use the word gene to refer to a distinct DNA segment with defined nucleotide sequences that can be transcribed into RNA, which in turn can be translated into proteins and bestows different traits on the host organism; Francis Crick named this scenario the central dogma of molecular biology (Crick 1970). Furthermore, since humans are a diploid species, each gene is transmitted as two copies, one from each parent, corresponding to the genotype (the sum of all genetic variations in an individual). Interactions between these copies (also sometimes referred to as alleles) and their expression give rise to the "*phenotype*", the different observable traits of an organism, from the morphology to physiological properties (Bateson and Saunders 1902). A classic example is eye color in humans; two copies of the allele corresponding to the blue color for the main gene involved are required to confer a blue phenotype (recessive alleles). In contrast, a combination of brown (dominant allele) and blue alleles will give rise to a brown phenotype. The transfer of these characteristics from parents to offspring is called heredity (Allen 2003). For most mammals, including humans, this process is performed through sexual reproduction, which is responsible for introducing a factor of chance into the process of heredity. As the alleles are transmitted equally but combine by chance, one cannot predict the phenotype of the offspring, only the statistical chance of any given combination. For example, parents with brown eyes can give birth to offspring with blue eyes, if both parents possess blue alleles, and if those blue alleles are transferred. The phenotype of the child depends on the specific genotype inherited, which is the consequence of a random combination of parental alleles. Notably, the botanist and monk Gregor Mendel originally studied and described the transmission frequencies of peas in the monastery garden, leading him to propose his three laws of inheritance in 1865 (Mendel 1866). Random combinations of alleles introduce variations between individuals, which are critical for the evolution of species and provide the basis of natural selection.

11.3. Random alterations

Interestingly, there is another source of variation inherent in all living systems. This one comes not from random combinations, but from random error or damage. "*Mutations*" correspond to changes in the DNA sequence altering the genotype, which can sometimes result in a change of the phenotype. As most of the genome is non-coding, most of the genetic alterations will actually remain neutral. But some can be beneficial and some will be negative, hence participating in the overall fitness of organisms (Kimura 1968). There are many sources of mutation. One comes from the fact that the DNA polymerases, the molecular machines that copy DNA templates to prepare for cell division, are themselves error prone; random genetic variation is, after all, a necessary perquisite for life on earth. But DNA mutations can also be caused by external damaging conditions. For example, DNA intercalating agents or UV emission from the sun can modify DNA at random loci. Fortunately, cells are equipped with machineries to detect and repair DNA errors, mutations and breaks; this field attracts much attention from cancer biologists (Sancar et al. 2004; Hakem 2008; Chatterjee and Walker 2017). As there are two DNA strands, limited damage to one strand can be repaired using the other strand as a template. Such events involve mismatch detection and excision repair of damaged DNA. But when both strands are damaged, repair becomes more complicated; cells use homologous recombination using sister chromatids, when possible, or homologous chromosomes as a template to repair the break. When there is no template available, such as before replication, a process called non-homologous end-joining will directly merge the two ends. But this process can lead to lost nucleotides, and non-matching ends can produce random insertions or translocations. Interestingly, this process opens up tremendous opportunities. Indeed, most of the genome editing techniques, including the infamous CRISPR-Cas9 technologies, rely on such random processes (Wang et al. 2016). Thus, chance plays an important role in creating genetic variation (recombination, mutation, deletions) that can drive evolution as well as cause disease.

11.4. Beyond the gene

The 20th century can be considered the century of the gene, from its conceptual invention to sophisticated genetic engineering (Miescher 1871; Johannsen 1909; Beadle and Tatum 1941; Pearson 2006). Major investments and technological breakthroughs drove two projects to sequence the human genome: the public sponsored, international Human Genome Project and the private Celera Genomics initiative (International Human Genome Sequencing Consortium 2001; Venter et al. 2001). The completion of the human genome sequencing project was described as the deciphering of the "Book of Life", renewing hopes to cure many genetic pathologies (Pennisi 2000). Indeed, if diseases arose from mutated genes, then being able to read and potentially rewrite the genome would create a boulevard of opportunities, from genetic prediction to genetic correction. The reality turned out to be more challenging. Only a tiny fraction of the human genome corresponds to actual genes, while the remaining parts are non-coding and intergenic. At first the parts of the genome that did not contain genes were referred to as "junk DNA", but it quickly became clear that these regions contained a critical piece of the puzzle and would provide clues to help understand how gene expression is regulated (Hong et al. 2016). Furthermore, the DNA molecule itself could not explain all heritable changes. The phenotype of the cell or the organisms does not solely rely on the gene sequence (genotype), but is also influenced by environmental factors and the cellular context. This helped fuel the emergence of the new field of epigenetics, which could help to explain how genes are regulated and would provide some clue to understanding unconventional, non-Mendelian inheritance that had stumped geneticists (Morange 2005). We suggest that, rather than emphasizing the confrontational or revolutionary nature of the genetics-epigenetics debate, it is more helpful to think of them as co-evolving.

Debates over the definition of the word "*epigenetics*" are almost as complicated as those attempting to discuss "*chance*" and there is much controversy (Ptashne 2007; Deans and Maggert 2015). Epigenetics can be considered as a recent field that takes its root from antiquity. Indeed, science has long been based on observation, and nature offers a lot of inspiration. Aristotle opened up fertilized eggs and proposed that embryonic development results from tissue interactions between the constituents of the embryo (Aristotle 350 BCE), coining the word "*epigenesis*". Epigenesis attracted support from philosophers, who placed it in opposition to preformationism or neo-formationism theories, which postulated that all organism features were already present in the embryo and that development was simply growth in size. The term epigenetics emerged in the mid-20th century, inspired by a fusion of epigenesis and genetics, whilst also suggesting features "on top of" genetics. Conrad H. Waddington coined this clever word, while building on previous developmental concepts, to define epigenetics as "*the branch of biology which studies the causal interactions between genes and their products, which bring the phenotype into being*" (Harvey 1651; Waddington 1942; Aristotle 350 BCE). Waddington introduced the metaphor of the "*epigenetic landscape*" to describe embryonic development, though much of the underlying biological mechanisms or concepts remained mysterious (Waddington 1957). Since then, many researchers have tried to refine epigenetic definitions and knowledge, including seminal contributions from Robin Holliday, Adrian Bird and Roberto Bonasio (Holliday 1994; Bird 2007; Bonasio et al. 2010). Although there are many definitions of epigenetics, one of the most commonly accepted ones is that by Riggs et al.: "*the study of mitotically and/or meiotically heritable changes in gene function that cannot be explained by changes in DNA sequence*" (Riggs et al. 1996).

11.5. Epigenetic variation

Ever since the first use of the word epigenetics over six decades ago, the scientific community has continued to plunge deeper and deeper into our understanding of DNA and inheritance (Olins and Olins 2003). We now know that DNA, the heritable genetic material, is made up of two anti-parallel strands that form a double helix (Hershey and Chase 1952; Franklin and Gosling 1953; Watson and Crick 1953; Wilkins and Randall 1953). For instance, the human genome contains two copies of around 3 billion DNA base pairs, arranged into 46 chromosomes per cell, corresponding to nearly 2-m long DNA. The packaging of this long molecule into this tiny nuclear space is achieved by chromatin, a versatile structure of compacted DNA, whose proper dynamics are essential for genome functions (Flemming 1882). DNA is not naked and free-floating, but is wrapped around nucleosomes, comprising an octamer with two copies each of the core histone proteins (H2A, H2B, H3 and H4). This chromatin structure is key for most epigenetic events and all associated contexts, from physiological processes to diseases (Misteli 2010; Reddy and Feinberg 2013; Lupianez *et al.* 2016). For example, some plants require a period of cold in order to flower, which usually occurs after the winter (Amasino 2004). This process, called vernalization, was elucidated in *Arabidopsis* species and requires the coordinated actions of proteins involved in chromatin organization, such as the Polycomb complex (Angel *et al.* 2011), histone variants (Deal *et al.* 2007) and lncRNAs (long non-coding RNA) (Heo and Sung 2011), all of which are critical elements for gene regulation. Despite its highly organized nature, we can legitimately ponder whether chance plays a role in chromatin behavior and epigenetic mechanisms.

The nuclear chromatin is organized at different levels and is highly dynamic throughout the cell cycle (Maze *et al.* 2014; Sitbon *et al.* 2017). During interphase, chromatin can be distinguished by two main categories inside the nucleus, originally defined by DNA staining as a function of the compaction level (Heitz 1928). These two states correspond to "*euchromatin*" and "*heterochromatin*". Heterochromatin refers to a DNA structure that does not change compaction status throughout the cell cycle, in contrast to dynamic euchromatin. In mitosis, when the cell is about to divide, chromatin is even more compacted in visible X-shaped chromosomes. Many chromosomal landmarks have been described, such as the end of each chromosome arm, which are protected by a structure called the telomere. The location of centromeres, however, is less clearly defined. Indeed, centromeres represent a specific structure responsible for ensuring chromosome segregation during cell division. In human cells, as in many plants and other metazoans, there is only one centromere per chromosome, whose location is mostly defined by the DNA sequence and consequent loading of a specific H3 histone variant (Earnshaw and Rothfield 1985; Palmer *et al.* 1991; Blower and Karpen 2001; Sullivan *et al.* 2001). But in contrast to monocentric chromosomes, the case of centromere location in holocentric chromosomes is more complicated (Drinnenberg *et al.* 2016). In such situations, the entire length of the chromosome can act as centromeres. There is no primary constriction, but a spreading over the whole chromosome. And this is far from being an exception; many species such as *Caenorhabditis elegans* or *Bombyx mori* possess such chromosomes. Interestingly, it is unknown whether their positions can be predicted or if they appear randomly. Thus, more insights need to be generated in order to decipher whether or not such centromere location is distributed by chance.

At the same level, another interesting event occurs regarding sex differences. In mammals, two distinct sex chromosomes establish the sex of the organism. With the exception of a few cases, a female has two X chromosomes (one inherited from her mother and the other from her father), while a male will have one X and one Y chromosome (the former from his mother and the latter from his father). Interestingly, two doses of the X chromosome is lethal for the organism (Weitzman 2006). This paradox concerning females has been solved by the discovery of an epigenetic mechanism inactivating one of the two X chromosomes (Lyon 1961; Galupa and Heard 2018). Such X chromosome inactivation seems to occur randomly, targeting in each cell of the organism either the maternal or paternal X chromosome. This phenomenon can be visualized with calico cats, since the color of their fur is encoded by genes on the X chromosome. Inactivation of one X will switch off the orange color, while the inactivation of the second will switch off the black color. This results in a beautiful pattern of black and orange patches. As for centromeres, sex determination can be different between species. In reptiles, it is mostly dependent of temperature during egg incubation (Charnier 1966; Bull and Vogt 1979; Ferguson and Joanen 1982; Pieau et al. 1999). For instance, below a certain temperature threshold, embryos inside the eggs will become males. The random location of each egg inside the nest will therefore be key for the sex determination, since the temperature varies at different positions. Of note, such dependency is regulated by proteins that play on epigenetics marks, by modulating the expression of the male sex-determining gene Dmrt1 of the red-eared slider turtle *Trachemys scripta elegans* (Ge et al. 2018). Thus, many epigenetic processes rely in part on chance, from alteration of gene expression to chromosome compaction.

11.6. Concluding remarks

Epigenetic mechanisms might help explain why monozygotic identical twins display different phenotypes. The chance interactions with environmental factors may affect epigenome structures, genomic functions and gene expression. Whether chance truly participates in biological processes or is a consequence of our misunderstanding of life is a key question. Some thinkers have described "chance" as all the phenomena that we do not know how to explain properly. This is reminiscent of some definitions of epigenetics, where students are advised that "if you don't know the cause, you say it's epigenetic" (Nat. Biotechnol. 2010). The mysteries and debates around chance and necessity encourage us to plunge deeper into our disciplines in the search for knowledge to explain the unknown. The interdisciplinary environment created at the Colloques de Cerisy provided a stimulating context to reflect on the role of chance, serendipity, accidents, hazards, unpredictability and luck in living systems. Rather than reaching a clear conclusion, we are left with much "food for thought" and we encourage our colleagues whose valuable work and contributions on this topic were not cited to remember the parting words of Louis Pasteur: *"Chance favors only the prepared mind"*.

11.7. Acknowledgments

We thank the organizers and all the participants of the Colloque de Cerisy *"Le hasard, le calcul et la vie"* (Chance, calculation and life) for inviting us to speak and for lively discussions. We also apologize to all colleagues whose valuable work and contributions on this topic were not cited due to space constraints. The Weitzman laboratory is supported by the LabEx "Who Am I?" #ANR-11-LABX-0071 and the Université de Paris IdEx #ANR-18-IDEX-0001 funded by the French Government through its "Investments for the Future" program and JBW, a Senior Member of the Institut Universitaire de France (IUF).

11.8. References

Allen, G.E. (2003). Mendel and modern genetics: The legacy for today. *Endeavour*, 27(2), 63–68.

Aristotle. (350 BCE). On the generation of animals.

Bateson, W. and Saunders, E.R. (1902). The facts of heredity in the light of Mendel's discovery. *Reports to the Evolution Committee of the Royal Society*, 126–160.

Beadle, G.W. and Tatum, E.L. (1941). Genetic control of biochemical reactions in Neurospora. *PNAS*, 27(11), 499–506.

Bird, A. (2007). Perceptions of epigenetics. *Nature*, 447(7143), 396–398.

Blower, M.D. and Karpen, G.H. (2001). The role of Drosophila CID in kinetochore formation, cell-cycle progression and heterochromatin interactions. *Nat. Cell Biol.*, 3(8), 730–739.

Bonasio, R., Tu, S., Reinberg, D. (2010). Molecular signals of epigenetic states. *Science*, 330(6004), 612–616.

Bull, J.J. and Vogt, R.C. (1979). Temperature-dependent sex determination in turtles. *Science*, 206(4423), 1186–1188.

Charnier, M. (1966). Action of temperature on the sex ratio in the Agama agama (Agamidae, Lacertilia) embryo. *C.R. Seances Soc. Biol. Fil.*, 160(3), 620–622.

Chatterjee, N. and Walker, G.C. (2017). Mechanisms of DNA damage, repair and mutagenesis. *Environ. Mol. Mutagen.*, 58(5), 235–263.

Crick, F. (1970). Central dogma of molecular biology. *Nature*, 227, 561–563.

Deans, C. and Maggert, K.A. (2015). What do you mean, "epigenetic"? *Genetics*, 199(4), 887–896.

Drinnenberg, I.A., Henikoff, S., Malik, H.S. (2016). Evolutionary turnover of kinetochore proteins: A ship of theseus? *Trends Cell Biol.*, 26(7), 498–510.

Earnshaw, W.C. and Rothfield, N. (1985). Identification of a family of human centromere proteins using autoimmune sera from patients with scleroderma. *Chromosoma*, 91(3–4), 313–321.

Ferguson, M.W. and Joanen, T. (1982). Temperature of egg incubation determines sex in *Alligator mississippiensis*. *Nature*, 296(5860), 850–853.

Flemming, W. (1882). Zellsubstanz, Kern und Zelltheilung. [Online]. Available at: https://openlibrary.org/books/OL20780192M/Zellsubstanz_Kern_und_Zelltheilung.

Fleming, A. (1929). On the antibacterial action of cultures of a penicillium, with special reference to their use in the isolation of *B. influenzæ*. *British Journal of Experimental Pathology*, 10(3), 226–236.

Franklin, R.E. and Gosling, R.G. (1953). Evidence for 2-chain helix in crystalline structure of sodium deoxyribonucleate. *Nature*, 172(4369), 156–157.

Galupa, R. and Heard, E. (2018). X-chromosome inactivation: A crossroads between chromosome architecture and gene regulation. *Annual Review of Genetics*, 52, 535–566.

Ge, C., Ye, J., Weber, C., Sun, W., Zhang, H., Zhou, Y., Cai, C., Qian, G., Capel, B. (2018). The histone demethylase KDM6B regulates temperature-dependent sex determination in a turtle species. *Science*, 360(6389), 645–648.

Hakem, R. (2008). DNA-damage repair; The good, the bad, and the ugly. *The EMBO Journal*, 27(4), 589–605.

Harvey, W. (1651). *Disputations Touching the Generation of Animals*.

Heitz, E. (1928). Das Heterochromatin der Moose. *Jahrbücher für Wissenschaftliche Botanik*, 69, 762–818.

Hershey, A.D. and Chase, M. (1952). Independent functions of viral protein and nucleic acid in growth of bacteriophage. *The Journal of General Physiology*, 36(1), 39–56.

Holliday, R. (1994). Epigenetics – An overview. *Developmental Genetics*, 15, 453–457.

Hong, E.L., Sloan, C.A., Chan, E.T., Davidson, J.M., Malladi, V.S., Strattan, J.S., Hitz, B.C., Gadbank, I., Narayanan, A.K., Ho, M., Lee, B.T., Rowe, L.D., Dreszer, T.R., Rose, G.R., Podduturi, N.R., Tanaka, F., Hilton, J.A. and Cherry, J.M. (2016). Principles of metadata organization at the ENCODE data coordination center. *Database (Oxford)*.

International Human Genome Sequencing Consortium (2001). Initial sequencing and analysis of the human genome. *Nature*, 409, 860–921.

Johannsen, W. (1909). Elemente der exakten Erblichkeitslehre. *Nature*, 81(424).

Kimura, M. (1968). Evolutionary rate at the molecular level. *Nature*, 217, 5129, 624–626.

Lupianez, D.G., Spielmann, M., Mundlos, S. (2016). Breaking TADs: How alterations of chromatin domains result in disease. *Trends in Genetics*, 32(4), 225–237.

Lyon, M.F. (1961). Gene action in the x-chromosome of the mouse. *Nature*, 190(22), 372–373.

Maze, I., Noh, K.M., Soshnev, A.A., Allis, C.D. (2014). Every amino acid matters: Essential contributions of histone variants to mammalian development and disease. *Nat. Rev. Genet.*, 15(4), 259–271.

Mendel, G. (1866). Versuche über Pflanzen-Hybriden. *Verhandlungen des Naturforschenden Vereines, Abhandlungern*, 4, 3–47.

Miescher, F. (1871). Ueber die chemische Zusammensetzung der Eiterzellen. *Medicinisch-chemische Untersuchungen*, 4, 441–460.

Misteli, T. (2010). Higher-order genome organization in human disease. *Cold Spring Harb. Perspect. Biol.*, 2(8).

Morange, M. (2005). Quelle place pour l'épigénétique ? *Med. Sci.*, 21(4), 367–369.

Nat. Biotechnol. (2010). Making a mark. *Nature Biotechnology*, 28(1031) [Online]. Available at: https://www.nature.com/articles/nbt1010-1031.

Olins, D.E. and Olins, A.L. (2003). Chromatin history: Our view from the bridge. *Nat. Rev. Mol. Cell Biol.*, 4, 809–814.

Palmer, D.K., O'Day, K., Trong, H.L., Charbonneau, H., Margolis, R.L. (1991). Purification of the centromere-specific protein CENP-A and demonstration that it is a distinctive histone. *Proc. Natl. Acad. Sci. USA*, 88(9), 3734–3738.

Pearson, H. (2006). Genetics: What is a gene? *Nature*, 441(7092), 398–401.

Pennisi, E. (2000). Human genome. Finally, the book of life and instructions for navigating it. *Science*, 288(5475), 2304–2307.

Pieau, C., Dorizzi, M., Richard-Mercier, N. (1999). Temperature-dependent sex determination and gonadal differentiation in reptiles. *Cell Mol. Life Sci.*, 55(6–7), 887–900.

Ptashne, M. (2007). On the use of the word "epigenetic". *Curr. Biol.*, 17(7), R233–R236.

Reddy, K.L. and Feinberg, A.P. (2013). Higher order chromatin organization in cancer. *Semin Cancer Biol.*, 23(2), 109–115.

Riggs, A.D., Martienssen, R.A., Russo, V.E.A. (1996). Introduction. In *Epigenetic Mechanisms of Gene Regulation*. Laboratory Press, Cold Spring Harbor, New York, USA.

Sancar, A., Lindsey-Boltz, L.A., Ünsal-Kaçmaz, K., Linn, S. (2004). Molecular mechanisms of mammalian DNA repair and the DNA damage checkpoints. *Annual Review of Biochemistry*, 73, 39–85.

Sitbon, D., Podsypanina, K., Yadav, T., Almouzni, G. (2017). Shaping chromatin in the nucleus: The bricks and the architects. *Cold Spring Harb. Symp. Quant. Biol.*, 82, 1–14.

Sullivan, B.A., Blower, M.D., Karpen, G.H. (2001). Determining centromere identity: Cyclical stories and forking paths. *Nat. Rev. Genet.*, 2(8), 584–596.

Venter, C.J., Adams, M.D., Myers, E.W., Li, P.W., Mural, R.J., Sutton, G.G., Smith, H.O., Yandell, M., Evans, C.A., Holt, R.A. (2001). The sequence of the human genome. *Science*, 291(5507), 1304–1351 [Online]. Available at: https://science.sciencemag.org/content/291/5507/1304.

Waddington, C.H. (1942). The epigenotype. *Endeavour*, 1, 18–20.

Waddington, C.H. (1957). *The Strategy of the Genes: A Discussion of Some Aspects of Theoretical Biology*. George Allen & Unwin, Ltd, London.

Walpole, H. (1754). Letter from Horace Walpole to Horace Mann. Yale University, New Haven, CT.

Wang, H., La Russa, M., Qi, L.S. (2016). CRISPR/Cas9 in genome editing and beyond. *Annual Review of Biochemistry*, 85, 227–264.

Watson, J.D. and Crick, F.H.C. (1953). A structure for deoxyribose nucleic acid. *Nature*, 171, 737–738

Weitzman, J.B. (2006). Getting the right dose of sex (chromosomes). *J. Biol.*, 5(1), 1.

Wilkins, M.H. and Randall, J.T. (1953). Crystallinity in sperm heads: Molecular structure of nucleoprotein *in vivo*. *Biochim. Biophys. Acta*, 10(1), 192–193.

12

When Acquired Characteristics Become Heritable: The Lesson of Genomes

Considered as a cause of biological evolution at the beginning of the 19th century, before being clearly refuted by Genetics, the idea of acquired inheritance reappears today in genomes. While the initial hypothesis subordinated genes to their functions, which the facts exclude, its avatar makes them the masters of processes, having the misleading appearance of finality while they are clearly random. These ideas, illustrated with the help of recent examples from genome analysis, are discussed in the context of molecular mechanisms ensuring the propagation and exchange of genetic material in a living world, where the universality of DNA is what allows the horizontal acquisition of foreign genes.

12.1. Introduction

It took a very long time to resolve the apparent contradiction between the intergenerational permanence of living organisms and changes in biological evolution. The beginnings of transformism in the Age of Enlightenment (developed mainly by Lamarck) imagined the gradual formation of new varieties and species through the succession of adaptive modifications which, over the generations, would become heritable through an entirely mysterious process. This process has never received any scientific confirmation. The inheritance of acquired characteristics by individual adaptation does not exist. The formerly described cases of adaptive traits that mimic heredity (Landman 1991) are epigenetic in nature. The transformist idea taken by Darwin half a century later, emphasizing the reproductive advantage of individuals carrying certain changes, still stated the appearance of new species through gradual changes from ancestral species by making the implicit assumption

Chapter written by Bernard DUJON.

that these changes were heritable. However, as we now know, only genetic *mutations* are heritable, other phenotypic variations between individuals of a same species, whether random (phenotypic noise) or induced by external conditions, are not.

This was only a few years before Mendel (1866) laid the mathematical bases for the inheritance of traits, precisely defining "factors" that eventually lead to *genes* and their *alleles* 30 years later (Dujon and Pelletier 2019). Hence, at the end of the 19th century, without yet understanding what these "factors" were made of, their rules of transmission to offspring agreed well with the germinal plasma theory developed by Weissmann, explaining the intergenerational permanence by the exclusively vertical transmission of characters. In this context, the concept of *mutation*, initially defined by de Vries, allowed nascent Genetics to explain all the hereditary changes observed within lineages, even if they were still far from clarifying their molecular bases. The rapid success of early Genetics led to the ignorance of any horizontal acquisition of new characteristics through exchanges between organisms of distinct kinship.

Ironically however, we owe the discovery that DNA is the material support of heredity to this phenomenon. As early as 1928, Griffiths observed that, following the simultaneous injection of living non-virulent pneumococci (type R) together with heat-killed virulent pneumococci (type S), mice died of pneumonia due to the proliferation of virulent S-type pneumococci. The factor released by the dead S cells was therefore capable of transforming the living R cells into S cells (Griffiths 1928). A few years later, after the biochemical purification of the products released by the dead S cells of pneumococci, Avery and his collaborators demonstrated that this factor was only DNA (Avery *et al.* 1944). The historical identification of the genetic material is therefore based on the phenomenon of horizontal transfer of genetic material from one line of dead bacteria to another line of living bacteria, a phenomenon which was supposed not to exist in nascent Genetics!

After elucidating the role of DNA, the initial developments of molecular biology continued to ignore horizontal genetic transfers. DNA molecules were considered to replicate identically, ensuring the vertical transfer of genetic information from generation to generation (which is indeed their main role), with the possibility of accidents during this process (explaining at least part of the mutations), but never receiving any information from the cell or from external sources. It was not until the discovery of *retroviruses* in the early 1970s that it was realized that the RNA molecules that they bring to the infected cell are copied into DNA, which is then inserted into the host's chromosomes. Therefore, that foreign genetic information could be acquired by a genome. This phenomenon is happening right now before our eyes, as shown by the fact that young Australian Koalas, *Phascolarctos cinereus*, are born carrying the genes of an Asian retrovirus, *KoRV*, in their genomes that they have never encountered, but which infected the germline of one of their recent ancestors (Greenwood *et al.* 2017). Therefore, through the process of *endogenization*, an infectious agent horizontally transmissible between genetically unrelated individuals became a new element of the genome, vertically transmissible between generations.

Beyond simple viruses, recent developments in Genomics demonstrate the evolutionary importance and generality of the horizontal transfer of genetic information between organisms of distinct lineages. Bacterial or fungal genes are found in the genomes of plants, animals or unicellular eukaryotes, eukaryotic genes in bacteria, plant genes in fungi, etc. Our own genome has dozens of genes acquired from other living species, not to mention from viruses. So, what do we know about the phenomenon of horizontal transfer of genetic information between organisms with no parental relationship? What are the molecular mechanisms involved? And what are the consequences?

12.2. Horizontal genetic exchange in prokaryotes

It is now clear that horizontal gene transfer plays a major role in the world of prokaryotes (bacteria and archaea) where they dramatically accelerate adaptation to the changing conditions that these organisms are continually facing. The development of resistance to antibiotics illustrates this phenomenon. Resistances selected in one species can quickly spread to other species. The intensity of horizontal genetic exchanges between prokaryotes is such that, in many bacterial species, the set of genes common to all members of the species (core-genome) is smaller than the set of genes that varies between individual genomes owing to horizontal exchanges (Hall *et al.* 2017).

Several mechanisms are responsible for horizontal gene transfer in prokaryotes. First, the *transformation* of bacteria by the DNA present in the medium. The generality of this mechanism, at the origin of the discovery of the role of DNA, as already mentioned, was recognized from the 1950s. The molecular details of the penetration of external DNA into cells can vary depending on the species but, in all cases, the DNA ends up integrating into the genome of the host cell, thus becoming transmissible to its descendants. At around the same time, the phenomena of bacterial *conjugation* and *lysogeny* were also discovered. Bacterial conjugation is the fact that some bacteria can transmit fragments of their chromosomes to other more or less related bacteria through simple contact. We now know that DNA travels from one bacterium to another through specialized protein tubes whose elements are encoded by plasmids carried by bacteria capable of conjugation. Bacterial lysogeny is due to the fact that certain bacterial viruses – the *temperate bacteriophages* – instead of multiplying immediately after infection, integrate their genome into that of the host cell where it becomes silent. The bacteria then become *lysogenic*, that is, able to perpetuate the silent viral genome for an unlimited number of generations without ever producing bacteriophages, unless special conditions accidentally reactivate the viral genome. In this case, a distant descending bacterium will produce numerous infectious bacteriophages, which lyse the cell, even though this cell has never been infected itself.

Another interesting example of the acquired inheritance of nucleic acid sequences in bacteria is provided by the CRISPR-Cas bacterial immunity system now widely used as a molecular genome editing tool (Makarova *et al.* 2015). In this case, a particular region of the bacterial chromosome made up of short palindromic repeat sequences serves as a reservoir for integrating short fragments of DNA from foreign viruses or plasmids which, after having entered the cell, have been cut out by the endonuclease Cas. These fragments are thus perpetuated identically from generation to generation by bacteria, providing them with traces of ancestral accidental encounters in the form of small RNAs, as the reservoir is transcribed and the long RNA produced is cut into small RNAs at the level of the palindromic repeats. If a foreign DNA that enters a bacterial cell of this line carries a sequence that matches one of these small RNAs, it will immediately be cleaved by the Cas endonuclease and destroyed.

It should be noted that, in all above-cited phenomena where acquired characters become heritable, the molecular mechanisms allowing the phenomenon are pre-existing to the acquisition events that, themselves, are based on random encounters (environmental DNA, bacteriophage, conjugation partner). This is a general characteristic of horizontal genetic transfers.

12.3. Two specificities of eukaryotes theoretically oppose horizontal gene transfer

It was long believed that, as opposed to prokaryotes, horizontal genetic transfers were by far more limited, if not non-existent, in eukaryotes for at least two reasons: intracellular compartmentalization and, for multicellular organisms, the differentiation of germ lines. Indeed, intracellular compartmentalization forces foreign DNA to find its way through a complex network of cytoplasmic membranes before it can reach the organelles containing the genome of the host cell: the nucleus and mitochondria and, if present, chloroplasts. The discovery of methods for transforming eukaryotic cells in the laboratory during the 1970s showed that this does not constitute a real obstacle to horizontal gene transfer. Many eukaryotic cells (yeasts, animal or plant cells *in vitro*, etc.) are capable of receiving foreign DNA (*transgenesis*) and transmitting it to their descendants, either in the form of an additional autonomous element (*episome*), or after integration into their chromosomes. Gene therapy is based on this principle.

However, for many multicellular eukaryotic organisms there is an additional barrier to the hereditary transmission of a foreign gene to the offspring: the transformed cell must be able to give rise to a whole new organism. For many plants, regeneration from callus is possible in the laboratory and this technique is at the origin of many transgenic lines. However, under natural conditions, many multicellular eukaryotes have obligatory sexual reproduction with germline differentiation. For a horizontal genetic transfer to become heritable, it therefore has to affect a germ cell which, as many examples show, is perfectly possible. However, before going any further, how can one be sure that a given genetic element arose through a phenomenon of horizontal transfer from another organism and not from the normal vertical inheritance within a lineage?

12.4. Criteria for genomic analysis

Beyond the experimental cases produced in the laboratory, this demonstration relies on the analysis of genomes, and therefore of DNA sequences. This demonstration becomes more and more difficult as the horizontal transfer event becomes more distant in time, leaving more possibilities during successive generations, for the foreign gene to adapt to its host genome by mutation and increase its chances of dissemination among descendants. The phenomenon of horizontal transfer is therefore globally underestimated, with a bias in favor of the most recent events and a difficulty in identifying the oldest ones, even if they have played a very important evolutionary role.

Three types of signatures can suggest the existence of a gene (or a group of genes) in a genome resulting from a horizontal transfer event: the existence of a DNA composition bias, the presence of an additional element within a segment of conserved synteny between related species and, above all, the phylogenetic discordance of the sequences. The first anomaly occurs when the base composition of the DNA of the donor organism differs sufficiently from that of the recipient organism, such that the segment of DNA of foreign origin is easily distinguishable from the host genome. It will be noted, however, that this criterion alone is not sufficient enough to demonstrate a horizontal transfer of genes, as interspecific hybridizations (see below) can lead to the same phenomenon. The second anomaly can be observed by comparing the genomes of species that are evolutionarily close enough to have retained long segments of chromosomes carrying the same genes in the same order (conservation of synteny). In this case, the presence of an additional gene or genetic element in one of the species compared to the others may suggest its acquisition through horizontal transfer, if this gene is not present elsewhere in the genomes of the other species (to eliminate the hypothesis of a simple displacement of the gene by chromosomal translocation). It will be noted however that, even in this case, this criterion alone is not sufficient enough to demonstrate a horizontal transfer, as the additional gene could have been lost in all of the other species (a hypothesis that is becoming less and less probable as the number of studied species increases).

Finally, it is the phylogenetic discordance criterion that is the most demonstrative. By comparing the suspected sequence of foreign origin with databases, one can find a greater similarity with sequences of distant species than with those of closely related species. This is how a very large number of examples of genes related to bacterial genes have been reported within eukaryotic genomes. By carefully examining their phylogenies, it is sometimes even possible to predict the precise origin of these genes (the particular bacterial group). Conversely, genes of eukaryotic origin are found in the genome of certain bacteria. In all cases, the strong similarity (sometimes quasi-identity) of a segment of DNA sequence between unrelated organisms is the best indication that a horizontal genetic transfer has taken place, which obviously favors the most recent events but also makes the identification of the phenomenon dependent on the content of the databases. Now, this content is very biased when compared to the actual diversity of the living world, contributing greatly to the overall underestimation of the importance of horizontal genetic exchanges.

12.5. Abundance of horizontal transfers in unicellular eukaryotes

According to the above criteria, traces of horizontal transfers are found in practically all of the genomes of unicellular organisms studied which belong to the different phylogenetic branches known in eukaryotes (Keeling and Palmer 2008). Several dozen, or even hundreds of genes per genome come from this mechanism in *Chromalveolata* (ciliates, brown algae, diatoms, *Plasmodium*, etc.), *Amoebozoa* (amoebae, etc.) or *Excavata* (*trypanosomes*, *leishmanias*, *trichomonas*, etc.). They are particularly numerous in parasites or in the ciliates of the rumen of ruminants, which coexist with an abundant microbial flora in an environment rich in DNA and carbohydrates (Ricard *et al.* 2006).

It is not always easy to determine the function of genes acquired through horizontal transfer in their new host. Some (the majority?) may have no action and are only temporarily present in the evolutionary lineage. However, in other cases the acquired genes confer a significant advantage to the recipient organisms through the immediate emergence of new functions. For example, in the genomes of certain yeasts capable of strict anaerobic fermentation such as *Saccharomyces*, we find a gene of bacterial origin, *URA1* for the dihydro-orotate dehydrogenase (DHODase), which is necessary for the biosynthesis of uracil. Yet, there was another gene, *URA9*, for the same enzyme in the ancestral yeasts (lost in *Saccharomyces*). While the ancestral eukaryotic enzyme requires atmospheric oxygen to function (excluding anaerobic life), its bacterial replacement does not. Here, the acquisition of the bacterial gene, probably from a *Lactococcus*, allowed some yeasts to live anaerobically (and therefore produce wine and beer!), while their ancestors were unable to. We can also mention that in certain yeasts, such as *Candida parapsilosis*, the acquisition of bacterial genes for a proline racemase gives them the capacity to metabolize the D isoschizomer of the amino, therefore, offering them a nutritional advantage (the acquisition of racemase genes is common in unicellular eukaryotes). We could multiply the examples. But, obviously, in unicellular organisms, the acquisition of a new function is directly transmissible to their descendants. What about multicellular eukaryotes with differentiated germ lines?

12.6. Remarkable horizontal genetic transfers in pluricellular eukaryotes

Remarkable examples show that the phenomenon of horizontal gene acquisition also exists in multicellular eukaryotic lines such as plants, animals, algae or fungi. The presence of bacterial genes in plant chromosomes is not limited to artificial genetically modified organisms (GMOs). Nature produces the equivalent. The sweet potato, for example, *Ipomoea batatas* a *Convolvulaceae* whose roots swell with reserves, carries a set of genes in its genome from a plasmid of the bacterium *Agrobacterium* (Kyndt *et al.* 2015). This bacterium is normally symbiotic with *Fabaceae* (peas, beans, soybeans, etc.) with which it forms atmospheric nitrogen fixation nodules. All varieties of natural or cultivated *I. batatas* studied (nearly 300 in total) carry *Agrobacterium* genes, but the ubiquity of certain genes in the cultivated varieties suggests that it is precisely these naturally acquired genes that gave the plant its useful characteristics for domestication.

Gene transfer to eukaryotes is not limited to the integration of bacterial genes. Fungal genes are found to be functional in animals. For example, while most aphids, such as *Acyrthosiphon pisum*, *Myzus persicae* or *Chaitophorus populifolli* (hemiptera insects), are green, some have red-orange bodies because they produce colored carotenoids (tolurene, lycopene, and α, β and γ carotenes) from the colorless plant carotenoids they consume. Examination of their genomes shows the presence of genes from a fungal origin whose products (phytoene desaturase, lycopene cyclase/phytoene synthase) are responsible for the synthesis of these colored derivatives (Internal Aphid Genomics Consortium 2010). These acquired genes behave like perfect Mendelian factors within these aphids. The same phenomenon is found in various other arthropods such as the midges *Mayetiola destructor*,

Asteromyia carbonifera (dipterous insects), or the spider mite, *Tetranychus urticae*, an arachnid. They may have inherited the same horizontal transfer events as aphids from their distant common ancestors or, more likely, this illustrates recurrent events of capture of the same types of fungal genes (Cobbs *et al.* 2013).

Conversely, fungal species that have acquired genetic material from other eukaryotic lineages can be found. For example, there are lines of filamentous fungi and yeasts capable of assimilating nitrates, like plants do, which is unusual in the fungal world. These fungal species carry groups of typical plant genes in their genomes producing the permease, nitrate reductase and nitrite reductase necessary for this metabolism. We could multiply the examples among large groups of multicellular eukaryotes (Soucy *et al.* 2015). Smelts received the AFP type II antifreeze protein gene (which allows them to frequent icy water) from another fish, such as herring (Graham *et al.* 2012). Insects survive the cyanide produced by the plants they feed on because of a detoxifying gene acquired from bacteria (Wybouw *et al.* 2014), etc. By what mechanisms do such genetic transfers take place?

12.7. Main mechanisms of horizontal genetic transfers

We know of the different processes by which DNA from one organism can enter the cells of another organism. By using the enzymes of the recipient cell, or by inducing the synthesis of specific enzymes, it can integrate into the host's DNA, thus becoming a permanent genetic element of the latter (Figure 12.1). Certain mechanisms such as transformation or viral infection exist in both prokaryotes and eukaryotes. Others, like conjugation, are specific to bacteria, while others, like endosymbiosis, mainly affect eukaryotic cells.

Long-lasting associations between different organisms, as in cases of parasitism, commensalism or symbiosis, promote interspecific exchanges of genetic material. Examples exist in all categories of organisms. *Wolbachia pipiensis* genes are found integrated into the genomes of almost all insects, arachnids or nematodes infected with this α-proteobacterium, which lives as an endosymbiont in their germ cells (Dunning Hotopp *et al.* 2007). In insects, the presence of *W. pipiensis*, transmitted to offspring through the maternal line, tends to increase fertility (Toomey *et al.* 2013). In these cases, there is a transfer of DNA from the endosymbiont to the nuclei of the host cells. In other cases, the transfer takes place in the reverse direction. For example, legionella, γ-proteobacteria, some of which are responsible for lung infections in humans, carry many signatures of eukaryotic genes acquired from their hosts (amoeba) in their genomes, within which they typically multiply (Gomez-Valero *et al.* 2019). The same is true of plant parasites of other plants (Clarke *et al.* 2019). For example, the genome of *Striga hermonthica*, a dicot in the group of broomrapes that parasitize tropical grasses, contains genes specific to monocots from which it extracts nutrients via the haustoria affixed to their roots (Yoshida *et al.* 2010).

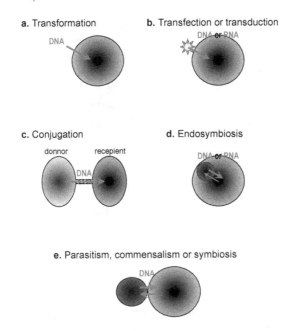

Figure 12.1. *The main types of gene transfer mechanisms. For a color version of this figure, see www.iste.co.uk/gaudin/chance.zip*

COMMENTS ON FIGURE 12.1.– *(a and b) A living prokaryotic or eukaryotic cell (gray circle with the dark central zone symbolizing the genome) can receive DNA from its immediate environment (in single-stranded or double-stranded form) by a transformation process, or receive a nucleic acid (DNA or RNA) via a virus. In the latter case, the nucleic acid can carry the genes of the virus alone or those of other organisms carried by the viruses from previous infectious cycles. (c) Certain bacterial cells (yellow oval) develop conjugation mechanisms during which they inject their DNA into another bacterial cell (gray oval), or sometimes into a eukaryotic cell, via specialized protein tubes (dotted lines) produced by the conjugative plasmids which they carry. (d) In eukaryotic cells, the presence of permanent endosymbionts resulting from phagocytoses or infections facilitates the transfer of nucleic acids between the endosymbiont (blue circle) and the host. Endosymbionts can be eukaryotic or prokaryotic and have widely varying degrees of host specificity. (e) Stable associations between distinct organisms also facilitate the transfer of DNA between cells in prolonged contact.*

Symbiotic interactions can also encompass complex populations of organisms, as in the case of intestinal flora (microbiota), within which many horizontal genetic exchanges are expected. The discovery that some Japanese people have acquired the capacity to digest seaweed porphyrins (polysaccharides that cannot be assimilated by human digestive enzymes) provides a spectacular example of this situation (Hehemann *et al.* 2010). Here, analysis shows that *Bacteroides* from the intestinal flora of regular sushi consumers have acquired β-porphyrinase and β-agarase genes through horizontal transfer from marine bacteria living on the surface of the red seaweed, *Porphyra*, used in the preparation of sushi. These genes are absent from *Bacteroides* of the intestinal flora in human populations that do not regularly consume this type of sushi. Horizontal genetic transfers therefore offer a very rapid mechanism of adaptation to environmental conditions.

Likewise, infectious agents with a broad host spectrum or mobile genetic elements can serve as vectors of gene exchange between organisms. For example, symbiotic viruses (*Bracovirus*) from wasps that lay their eggs in butterfly caterpillars have been shown to be vectors of gene transfer to lepidopteran insects (Drezen *et al.* 2017). The same is true of *Baculoviruses* which infect many lineages of insects. Even without carrying foreign genes, viruses can themselves be a source of acquisition for new functions by multicellular eukaryotes. As an example, the case of the mammalian placenta is spectacular. Mammalian genes that direct the synthesis of syncytins – the proteins that interface fetal cells with maternal cells and, for some of them, suppress the immune rejection response – are derived from genes of retrovirus encoding the envelop the protein that ensures their adhesion to the cells they infect (Lavialle *et al.* 2013). Our viviparity is therefore the result of the viral infection of our very distant ancestors during which, as in the case of today's Koalas, retroviruses have succeeded in integrating their genome into the germ cells of their host. The analysis of the genomes of the various orders of mammals shows that the phenomenon occurred many times during evolution, leading to different syncytins – and therefore different forms of placentas – after an initial event at the origin of placental mammals (marsupials and eutherians), while monotreme mammals remained oviparous. The fact that the only known viviparous lizards, *Mabuya sp.*, have developed a placenta from the same retroviral genes (Cornelis *et al.* 2017) indicates the recurrence of this type of horizontal acquisition and shows that a major evolutionary innovation such as viviparity can be the result of random accidents like the encounter of oviparous vertebrates with viruses.

The ability of different organisms to acquire foreign genes is not equivalent. Primates, including humans, show a more limited acquisition rate than insects or nematodes, if one judges by the number of genes acquired through horizontal transfers found in their genomes. In this respect, it is interesting to note that organisms which are able to survive extreme conditions, such as tardigrades (Boothby *et al.* 2015), or bdelloïd rotifers, which have lost sexual reproduction

(Flot *et al.* 2013), show very high numbers of foreign genes in their genomes acquired horizontally from a wide variety of different sources.

12.8. Introgressions and limits to the concept of species

Genetic exchanges between members of distinct species is not limited to the horizontal transfers described above. The progress of Genomics and the enrichment of databases show us that it is not uncommon to find some segments (which can carry many genes) in a genome, whose sequences indicate that they come from other organisms. This phenomenon concerns related species considered as distinct, but which can actually be quite distant in evolution from their last common ancestor.

Genomes of the holarctic yeasts *Saccharomyces uvarum*, for example, often show DNA segments of several dozens (or even hundreds) of kilobases originating from other species such as *S. eubayanus* or *S. kudriavzevii*, whose genomes are more distant than that of humans with various species of primates (Albertin *et al.* 2018). Likewise, in the genomes of various wine isolates of *S. cerevisiae*, we find long DNA segments originating from far more distantly related yeasts such as *Zygosaccharomyces baillii* (Galeote *et al.* 2011) or *Torulaspora microellipsoides* (Marsit *et al.* 2015), whose sequence divergences with *S. cerevisiae* exceed those between mice and humans. These gene transfers do not result from the mechanisms described above. They are explained by a mechanism of genetic *introgression*, often used by producers of new plant or animal varieties. In this scheme, a fertile hybrid, resulting from an interspecific cross (a case more frequent than usually imagined), is backcrossed with the species which we wish to improve to produce, after a succession of backcrosses of the progeny, descendants carrying only a few fragments from the other species (those carrying the genes selected by the producer) in their genomes. The mechanism of loss of heterozygosity, observed in the clonal progenies of hybrid yeast genomes (Dujon and Louis 2017), is prone to accelerate the phenomenon of introgression by aleviating the need for multiple backcrosses. In the fungal world, the temporary fusion between cells also probably facilitates the genetic introgression between distant species.

But more spectacularly, the same phenomenon is observed in animal species with obligatory sexual reproduction, such as mammals. Like in yeasts, the genomes of lions (*Panthera leo*), leopards (*P. pardus*) and jaguars (*P. onca*), three species easily recognizable from one another, have exchanged long segments of chromosomes during their recent evolution, and these introgressions gave rise to phenomena of intraspecific positive selections (Figueiró *et al.* 2017). The presence of segments of chromosomes of Neanderthal or Denisovan origin in modern human populations belongs to the same mechanism (Paabo 2015). The fact that, in felines, the presence of chromosome segments of foreign origin is enriched in chromosomal regions with a high rate of recombination (Li *et al.* 2019) clearly shows the effect of existing molecular mechanisms on the rates of horizontal genetic exchange between distinct lineages.

12.9. Conclusion

Beyond the ironic fact that the historic discovery of the role of DNA was based on a mechanism ignored by nascent Genetics – because superficially confused with old and false ideas – the recent recognition by Genomics of the generality of the horizontal transfer of genetic material sheds light on a previously missing part in our understanding of biological evolution: the importance of random qualitative jumps in addition to the gradual adaptations submitted to selection. Obviously, not all horizontal genetic acquisition events lead to notable evolutionary advances – the vast majority probably disappears very quickly – but the small number of those that leave long lasting traces in genomes is sufficient to highlight the importance of the phenomenon. In doing so, horizontal transfers reinforce the random dimension of evolution, but in a way that increases the role played by ecosystems because, if the molecular mechanisms are pre-existing (enzymes for transferring and modifying nucleic acids), the nature of the genetic elements acquired depend on encounters within the biosphere, which are fortuitous but not equiprobable.

The inheritance of acquired traits as envisioned by the pioneers of transformism does not exist, but traits acquired by accident can become heritable because DNA, the element of uniqueness of the living world, abounds in the environment. Moreover, Genomics shows us that the boundaries between species are much less tight than generally accepted. Sexual reproduction, a regular mechanism of genetic exchange between members of a same species at each generation, also contributes to the formation of chimeric genomes between members of more or less related species.

12.10. References

Albertin, W., Chernova, M., Durrens, P., Guichoux, E., Sherman, D.J., Masneuf-Pomarede, I., Marullo, P. (2018). Many interspecific chromosomal introgressions are highly prevalent in holarctic *Saccharomyces uvarum* strains found in human-related fermentations. *Yeast*, 35, 141–156.

Avery, O.T., MacLeod, C.M., McCarty, M. (1944). Studies on the chemical nature of the substance inducing transformation of pneumococcal types. Induction of transformation by a desoxyribonucleic acid fraction isolated from *Pneumococcus* type III. *J. Exp. Medicine*, 79, 137–158.

Boothby, T.C. (2015). Evidence for extensive horizontal gene transfer from the draft genome of a tardigrade. *Proc. Natl. Acad. Sci. USA*, 112, 15976–15981.

Clark, C.R., Timko, M.P., Yoder, J.I., Axtell, M.J., Westwood, J.H. (2019). Molecular dialog between parasitic plants and their hosts. *Annual Rev. Phytopathol.*, 57, 13.1–13.21.

Cobbs, C., Heath, J., Stireman 3rd J.O., Abbott, P. (2013). Carotenoids in unexpected places: Gall midges, lateral gene transfer, and carotenoid biosynthesis in animals. *Molecular Phylogenetics and Evolution*, 68, 221–228.

Cornelis, G., Funk, M., Vernochet, C., Leal, F., Tarazona, O.A., Meurice, G,. Heidmann, O., Dupressoir, A., Miralles, A., Ramirez-Pinilla, M.P., Heidmann, T. (2017). An endogenous retroviral envelope syncytin and its cognate receptor identified in the viviparous placental *Mabuya* lizard. *Proc. Natl. Acad. Sci. USA*, 114, E10991–E11000.

Drezen, J.M., Josse, T., Bézier, A., Gauthier, J., Huguet, E., Herniou, E.A. (2017). Impact of lateral transfers on the genomes of *Lepidoptera*. *Genes*, 8(11), E315.

Dujon, B.A. and Louis, E.J. (2017). Genome diversity and evolution in the budding yeasts (*Saccha-romycotina*). *Genetics*, 206, 717–750.

Dujon, B. and Pelletier, G. (2019). *Trajectories of Genetics*. ISTE Ltd, London, and John Wiley & Sons, New York.

Dunning Hotopp, J.C., Clark, M.E., Oliveira, D.C., Foster, J.M., Fischer, P., Muñoz Torres M.C., Giebel J.D., Kumar, N., Ishmael, N., Wang, S., Ingram, J., Nene R.V., Shepard, J., Tomkins, J., Richards, S., Spiro D.J., Ghedin, E., Slatko B.E., Tettelin, H., Werren J.H. (2007). Widespread lateral gene transfer from intracellular bacteria to multicellular eukaryotes. *Science*, 317, 1753–1755.

Figueiró, H.V., Li G., Trindade, F.J. *et al.* (2017). Genome-wide signatures of complex introgression and adaptive evolution in the big cats. *Science Adv.*, 3, e1700299.

Flot, J.F., Hespeels, B., Li, X. *et al.* (2013). Genomic evidence for ameiotic evolution in the bdelloid rotifer *Adineta vaga*. *Nature*, 500, 453–457.

Galeote, V., Bigey, F., Beyne, E., Novo, M., Legras J.L., Casaregola, S., Dequin S. (2011). Amplification of a *Zygosaccharomyces bailii* DNA segment in wine yeast genomes by extrachromosomal circular DNA formation. *PLoS One*, 6, e17872.

Gomez-Valero, L., Rusniok, C., Carson, D., Mondino, S., Pérez-Cobas, A.E., Rolando, M., Pasricha, S., Reuter, S., Demirtas, J., Crumbach, J., Descorps-Declere, S., Hartland, E.L., Jarraud, S., Dougan, G., Schroeder, G.N., Frankel, G., Buchrieser, C. (2019). More than 18,000 effectors in the *Legionella* genus genome provide multiple, independent combinations for replication in human cells. *Proc. Natl. Acad. Sci. USA*, 116, 2265–2273.

Graham, L.A., Li, J., Davidson, W.S., Davies, P.L. (2012). Smelt was the likely beneficiary of an antifreeze gene laterally transferred between fishes. *BMC Evol. Biol.*, 12, 190.

Greenwood, A.D. (2017). Transmission, evolution, and endogenization: Lessons learned from recent retroviral invasions. *Micr. Mol. Biol. Rev.*, 82(1), e00044-17.

Griffiths, F. (1928). The significance of pneumococcal types. *J. Hygiene*, 27, 113–159.

Hall, J.P.J. (2017). Sampling the mobile gene pool: Innovation via horizontal gene transfer in bacteria. *Phil. Trans. R. Soc. B*, 372, 20160424.

Hehemann, J.-H., Correc, G., Barbeyron, T., Helbert, W., Czjzek, M., Michel, G. (2010). Transfer of carbohydrate-active enzymes from marine bacteria to Japanese gut microbiota. *Nature*, 464, 908–912.

International Aphid Genomics Consortium (2010). Genome sequence of the pea aphid *Acyrthosiphon pisum*. *PLoS Biol.*, 8(2), e1000313.

Keeling, P.J. and Palmer, J.D. (2008). Horizontal gene transfer in eukaryotic evolution. *Nature Rev. Genetics*, 9, 605–618.

Kyndt, T., Quispe, D., Zhai, H., Jarret, R., Ghislain, M., Liu, Q., Gheysen, G., Kreuze, J.F. (2015). The genome of cultivated sweet potato contains *Agrobacterium* T-DNAs with expressed genes: An example of a naturally transgenic food crop. *Proc. Natl. Acad. Sci. USA*, 112, 5844–5849.

Landman, O.E. (1991). The inheritance of acquired characteristics. *Ann. Rev. Genetics*, 25, 1–20.

Lavialle, C., Cornelis, G., Dupressoir, A., Esnault, C., Heidmann, O., Vernochet, C., Heidmann T. (2013). Paleovirology of "syncytins", retroviral *env* genes exapted for a role in placentation. *Phil. Trans. R. Soc. B*, 368, 20120507.

Li, G., Figueiró, H.V., Eizirik, E., Murphy, W.J. (2019). Recombination-aware phylogenomics reveals the structured genomic landscape of hybridizing cat species. *Mol. Biol. Evol.*, 36, 2111–2126.

Makarova, K.S., Wolf, Y.I., Alkhnbashi, O.S. *et al.* (2015). An updated evolutionary classification of CRISPR-Cas systems. *Nature Rev. Microbiol.*, 13, 722–736.

Marsit, S., Mena, A., Bigey, F., Sauvage, F.X., Couloux, A., Guy, J., Legras, J.L., Barrio, E., Dequin, S., Galeote, V. (2015). Evolutionary advantage conferred by an eukaryote-to-eukaryote gene transfer event in wine yeasts. *Mol. Biol. Evol.*, 32, 1695–1707.

Mendel, G. (1866). Versuche über Pflanzen-Hybriden. *Verhandlungen des naturforshenden Vereines in Brünn*, 4, 3–47.

Paabo, S. (2015). *Néandertal, à la recherche des génomes perdus*. Edition Babel, Paris.

Ricard, G., McEwan, N.R., Dutilh, B.E., Jouany, J.P., Macheboeuf, D., Mitsumori, M., McIntosh, F.M., Michalowski, T., Nagamine, T., Nelson, N., Newbold, C.J., Nsabimana, E,. Takenaka, A., Thomas, N.A., Ushida, K., Hackstein, J.H., Huynen, M.A. (2006). Horizontal gene transfer from Bacteria to rumen Ciliates indicates adaptation to their anaerobic, carbohydrates-rich environment. *BMC Genomics*, 7, 22.

Soucy, S.M., Huang, J., Gogarten, J.P. (2015). Horizontal gene transfer: Building the web of life. *Nature Rev. Genet.*, 16, 472–482.

Toomey, M.E., Panaram, K., Fast, E.M., Beatty, C., Frydman, H.M. (2013). Evolutionarily conserved Wolbachia-encoded factors control pattern of stem-cell niche tropism in Drosophila ovaries and favor infection. *Proc. Natl. Acad. Sc. USA*, 110, 10788–10793.

Yoshida, S., Maruyama, S., Nozaki, H., Shirasu, K. (2010). Horizontal gene transfer by the parasitic plant *Striga hermonthica*. *Science*, 328, 1128.

Wybouw, N., Dermauw, W., Tirry, L., Stevens, C., Grbić, M., Feyereisen, R., Van Leeuwen, T. (2014). A gene horizontally transferred from bacteria protects arthropods from host plant cyanide poisoning. *eLife*, 3, e02365.

The Evolutionary Trajectories of Organisms are Not Stochastic

13.1. Evolution and stochasticity: a few metaphors

The theme of stochasticity in biological evolution is powered by metaphors or representations familiar to Western scientific cultures. The most common is undoubtedly the random nature of genetic mutations. In neo-Darwinian theory, these mutations are considered to be the substrate on which evolution occurs and upon which natural selection can act. This metaphor is associated with the common and rather inaccurate idea that events at the molecular level are naturally largely random. As if Brownian motion perpetuated itself up to the scale of biological macromolecules and made the events of mutations stochastic! To many, this presumptive randomness at the base of mechanisms for the generation of diversity suggests that evolution is a globally stochastic process.

Intellectually, this vision sits quite comfortably as one may naively think that it is opposed to extreme deterministic conceptions that are deemed outdated or not all that scientific. What could be better than chance ... and necessity (Monod 1970; Jacob 1977) to oppose those old or naive theories that favor mysterious or irrational determinisms (vitalism, predeterminism, even creationism, etc.)? Understanding the molecular support of inheritance and evolution has indeed led us to portray evolution as a stochastic process through the lens of modern scientific culture.

More seriously and theoretically, the neutral theory of evolution has provided a statistical framework for inferring whether genetic changes do in fact drift within populations as a consequence of mutation events (Kimura 1968). In this fundamental framework, it is the fixation of alleles through "genetic drift" (in the absence of the

Chapter written by Philippe GRANDCOLAS.

role of selection), which is *possibly* random *in fine*, and not so much the mutation events.

This first metaphor, making room for randomness, is in apparent agreement with the second metaphor, which concerns the presumptive non-repeatability of evolution. Clearly, the notion of repeatability would suggest that evolution is capable of producing similar results repeatedly under specific constraints. If this were the case, the part played by randomness in evolution would be reduced. The question on the repeatability of evolution was pleasantly discussed by Stephen Jay Gould who concluded that it would not be possible. As a result, this second metaphor has acquired considerable notoriety.

Finally, there is also a notion of chance in the occurrence of ecological collapse and biodiversity extinction events. This is linked to the image that has become symbolic of a large asteroid hitting the Earth and causing a mass extinction at the boundary between the Cretaceous and Tertiary periods. Stochasticity is here exemplified by cosmic trajectories and is not specific to life but nevertheless influences it in a significant way.

From the beginning, we see a difference in scale between these three metaphors; the first two are intrinsic to life either at the molecular or organismic level, while the third is extrinsic to life and operates at the cosmic level. This nestedness of scales from the molecule to the asteroids has an obvious consequence, namely that the most inclusive scale will strongly condition the final nature of evolution despite its characteristics at different nested scales. I will therefore endeavor to comment on the last two metaphors, determining on principle, with regard to the stochastic character of evolution.

13.2. The Gouldian metaphor of the "replay" of evolution

Stephen Jay Gould, a brilliant scientist and writer, has contributed much to the discussion around the question of determinisms in evolution. He thus helped understand how organisms are constructed during evolution, under the joint influence of mutations, natural selection and the existing and inherited states and structures of organisms at a given time of their evolution (Gould 2002). One of his texts, among many, had a great impact in its consideration of the probability that an organism would evolve the same way, by imagining if one could replay the film of Life, as if rewinding a tape cassette and replaying a film. This is the metaphor of the "replay" (Gould 1989). Gould's conclusion is that the evolution of organisms would certainly take a very different path if given the opportunity to start over at the previous stage.

If one could repeat the evolution with the same starting situations and if it led to very different results, as Gould postulates, this would suggest that the role of stochasticity is potentially important. Consequently, this would also suggest that developmental, genetic, structural, functional constraints, etc., which could channel evolution and make it take the same path repeatedly, are relatively weak.

How can we verify this proposition? The path taken by evolution can be analyzed through the study of phenotypes and their genetic architecture. Natural selection acts on the organism as a whole, in individuals whose phenotypes have been constructed by integrating genetics, "associated" organisms, development, and behavior. Natural (or sexual) selection is a "filter" whose strength can vary on these phenotypes: any differential reproduction can thus give rise to selection and each trait therefore has a potential contribution to fitness. Just like the stability of the phenotype or the capacity to evolve: canalization, plasticity, developmental stability, but also evolvability versus robustness can all be subject to selection.

To study the repeatability of evolution, we have two possibilities. On the one hand, comparative biology allows us to retrace what happened during evolution with the reconstruction of evolutionary relationships between organisms – phylogenies – on which any phenotypic or genetic change can be traced back. On the other hand, we can conduct repeatable evolutionary experiments on organisms which reproduce and develop very quickly: we cultivate these organisms several times over several generations, we subject them to the same selection pressures and we observe what evolutionary course took place each time.

13.3. The replay of evolution: what happened

All phylogenetic studies show the occurrence of many convergence events: different organisms in similar environments have evolved into similar phenotypes. Presumably similar selection pressures in different locations have led to similar phenotypes or to phenotypes that appear slightly different but with similar functions. Classically, many treatises on evolution list countless examples of truly striking convergence, such as the convergence between the South American armadillo and the pangolin in the old world, between the canine wolf and the Tasmanian marsupial wolf, between the flying squirrel and the phalanger, between the hedgehog and the echidna, etc.

Armadillos have bony dermal plates while pangolins have horny keratin scales. We will also find in other groups of organisms quite striking convergences within these mammalian morphologies, such as the millipedes Glomeridae or the insects Cockroaches Perisphaeriinae (*Pseudoglomeris*) whose external skeletons remarkably resemble that of armadillos. Can these similar structures be characterized as a case of convergence and the same evolutionary production or on the contrary as the occurrence of several different evolutionary paths? In the first case, evolution would indeed have been replayed, while in the second case, Gould's opinion is credited that evolution would never take exactly the same path if it could happen twice. It all depends on the scale of observation one takes. If we take a superficial look at these phenotypes, we will be tempted to answer "replay"; if we look closely, we can argue that the path is significantly different. This is also the reason why scientists have distinguished convergence from parallelism (Desutter-Grandcolas *et al.* 2005). Convergence can be distinguished *a priori* from any phylogenetic analysis: everyone can therefore see that the armadillo carapace is a different structure from the pangolin scales, not to mention the case of Arthropods and their exoskeleton, even if the resemblance is striking. The parallelism is more subtle and only a phylogenetic analysis shows *a posteriori* that the structures in question come from several evolutionary events. Parallelism would therefore constitute an undeniable argument for the repeatability of evolution, even greater than convergence.

Figure 13.1. *A famous representation of evolutionary convergence within mammals (Bourlière 1973)*

That being said, convergence (or parallelism) is very present in Life, to the point that phylogeneticists have worried about it regarding the validation of the quality of their reconstructions. Indeed, if convergence were so ubiquitous in organisms that it concerns innumerable characters, it could bias phylogenetic reconstructions by falsely taking as related those organisms that resemble each other by convergence more than by common inheritance. To take up our example again, armadillos and pangolins would have been falsely brought together as close relatives, although they belong to respectively distant branches (Cingulata [ex-Xenarthres] and Pholidotes) of the Mammalian tree. In reality, the convergence is seldom perfect enough to bias the phylogenies, except if there was a case for a great number of undetectable parallelisms, a situation very rarely encountered if ever. Armadillos and pangolins have never been considered close relatives...

This problem remains more important at the molecular level. The genetic code is simple, and each site on a DNA strand may ultimately take on a few states (with nucleotides A, G, C, T). If the mutation rate is high, a site can change several times and nothing will distinguish, for example, the original adenine from an adenine installed following two successive mutations (for example, A- > T then T- > AT).

Indeed, nothing looks more like an adenine or a thymine than any other adenine or a thymine on a DNA strand, regardless of their respective origins. In the context of phylogenetic reconstruction, we then term this site as being "saturated".

If we go back to the phenotypic scale, when we look at what has happened in most organisms, we therefore find empirically that evolution seems to be repeating itself, contrary to Gould's assertions. Nevertheless, it must be recognized that this pseudorepetition takes place from different points of origin (different organisms in fact) and that many of the examples cited concern more convergences – "approximate replays" – than parallelisms ("perfect replays"). Nevertheless, this suggests that the role of natural selection is important and that, from different situations, similar character states are still selected. This also suggests that this role exceeds that of stochasticity, with organisms and selection playing in concert toward often repeated results.

However, natural selection is not necessarily the only constraint leading to convergence. In other words, convergence is not only adaptive. Very often, phenotypes are not constructed with only direct and unequivocal correspondences between genes, phenotypes and environments, but are also constrained by their ontogenetic construction rules during development. Gould also discussed at length this developmental dimension in phenotype construction: the organism does not wipe out the structures and developmental patterns of its ancestors at each generation. Some phenotypic characters will therefore tend to appear several times during evolution for genetic or developmental reasons. All of these constraints mean that phenotypic evolution is therefore not a kind of lottery, each generation rolling the dice for its characteristics. As a matter of consequence, this allows the occurrence of complex evolutionary histories, even regressive ones, in evolutionary lineages. We can thus identify the four members of tetrapods, even in the vestigial state of mammals that have since become marine animals. Phylogenetic systematics makes it possible to understand how the hind limbs of the latter have regressed.

13.4. Evolutionary replay experiments

Another methodological step has been reached to deal with this question and to overcome the idiosyncratic bias of starting situations through the design of evolutionary experiments. Researchers selected organisms that reproduced and grew very quickly, and subjected them to calibrated selection pressures. After many generations under controlled conditions, it is then possible to compare the evolutionary paths taken by different experimental populations subjected to the same conditions.

Three famous experiments among others can be mentioned: 60,000 generations of bacteria (Lemski and Travisano 1994); 100 generations of drosophiles (Rose 1984) and 65 generations of mice (Garland *et al.* 2002).

In the particularly impressive experiment by Lenski and Travisano (1994), more than 60,000 generations were produced and samples from each of them frozen, making it possible to compare ancestral and descendant populations, showing the evolution of fitness after selection on a particular diet. The original 12 populations roughly converge in terms of fitness and some phenotypic traits, yet also show subtle differences.

A very rapid assessment of the first years of research on this subject in several laboratories brings the following results: evolution turns out to show a good number of parallelisms, but more or less heterogeneous depending on the case, similar genotypes with similar phenotypes, similar genotypes with phenotypic variations and similar evolution in fitness. If we implement successive stages of the experiment in different environments, we sometimes obtain the loss of the heterogeneity existing at the previous stage of the experiment. In some experiments, we also observe some interesting convergences on complex traits: for example, guppies evolve slower life histories without predation pressure; lizards evolve shorter limbs in narrow environments.

These evolutionary experiments are fascinating from the point of view of their analytical power due to the control over the starting state of organisms and over environmental conditions and pressures. However, they have not lasted long enough and one might wonder if the results obtained would remain similar after tens of thousands of additional generations.

13.5. Phylogenies versus experiments

Each method has its strengths and weaknesses. Comparative studies are of great temporal depth, involve many different groups of organisms and can still be conducted in a statistical context to avoid confirmation bias. On the other hand, of course, they cannot control all the starting conditions of these evolving situations. The experiments of evolution have opposite characteristics and their main limitation is their short time extent.

Both of these methods nevertheless show that convergence is pervasive, whether it is detected in the past or observed in the present. Obviously, these convergences and even these parallelisms are linked at least in some cases to a genetic architecture inherited from a common ancestor allowing for the recurrent appearance of similar characteristics, developmental constraints, and moreover similar selection pressures.

13.6. Stochasticity, evolution and extinction

The evolutionary trajectories of organisms are also strongly constrained by the environment including geography, climate and other interacting organisms. Living beings obviously very strongly depend on the environment in which they evolve. And this environment varies over time, fairly predictably (continental drift, cosmic cycles of energy or planetary orbit) or less predictably (falling asteroids).

We can thus show that the events of evolutionary diversification in very diverse organisms are conditioned over time by island biogeography: for example, when an oceanic island emerges, it allows speciation to operate within a well-defined time window (Nattier *et al.* 2017).

However, evolution is not only diversification but it is also extinction, even if this phenomenon is most often mistakenly regarded in the negative as a cause of loss and absence. Extinction is not necessarily brought about by an intrinsic component (maladaptation, genetic burden, etc.) but also by sudden extrinsic events (cosmic, volcanic, tectonic, etc.). And these extrinsic events can then lead to stochasticity in evolution. Entire clades can quickly disappear or shrink, allowing others to develop in the millions of years that follow, as happened in the aftermath of the KT crisis caused by the fall of a giant asteroid, an extrinsic event if ever there was! This great planetary crisis thus had innumerable evolutionary consequences for many different organisms. Everyone remembers the gradual extinction of the dinosaurs, of which only the birds (since re-diversified) remain, also making room for other groups of vertebrates, the mammals in particular.

13.7. Conclusion

From this very brief and very simple depiction of evolution, we can draw some conclusions concerning the question of the stochasticity of Life. The "replays" of evolutionary events show both contingency (in Gould's sense) and randomness, but also many repetitions. Obviously, repeated evolutionary events occurred for various intrinsic reasons (genetic architecture, developmental pathways, etc.) but also because different environments may sometimes generate comparable selection pressures. We must not forget that organisms evolve by themselves but under the pressure of the environment; extrinsic events can introduce a lot of "noise" in terms of mass extinctions and evolutionary opportunities.

Life evolution is therefore far from being completely random; on the contrary, it is limited, constrained and influenced by many factors, which make it partly repetitive but also very innovative. The notions of pre-adaptation/exaptation (Cuénot 1914; Gould and Vrba 1982) or tinkering and evolution (Jacob 1977) bear witness to this, suggesting that the evolution of organisms leads to the modification of various ancestral solutions rather than the invention *de novo* of classically optimal forms that would either all be similar to each other or completely different.

13.8. References

Bourlière, F. (1973). The comparative ecology of rain forest mammals in Africa and tropical America: Some introductory remarks. In *Tropical Forest Ecosystems in Africa and South America: A Comparative Review*, Meggers, B.J., Ayensu, B.J., Duckworth, E.S. (eds). Smithsonian Institution Press, Washington.

Cuénot, L. (1914). Théorie de la préadaptation. *Scientia*, 16, 60–73.

Desutter-Grandcolas, L., Legendre, F., Grandcolas, P., Robillard, T., Murienne J. (2005). Convergence and parallelism: Is a new life ahead of old concepts? *Cladistics*, 21, 51–61.

Garland, T., Morgan, M.T., Swallow, J.G., Rhodes, J.S., Girard, I., Belter, J.G., Carter, P.A. (2002). Evolution of a small-muscle polymorphism in lines of house mice selected for high activity levels. *Evolution*, 56, 1267–1275.

Gould, S.J. (1989). *Wonderful Life: The Burgess Shale and the Nature of History*. Norton, New York.

Gould, S.J. (2002). *The Structure of Evolutionary Theory*. The Belknap Press of Harvard University Press, Cambridge.

Gould, S.J. and Vrba, E.S. (1982). Exaptation – A missing term in the science of form. *Paleobiology*, 8, 4–15.

Jacob, F. (1977). Evolution and tinkering. *Science*, 196, 1161–1166.

Kimura, M. (1968). Evolutionary rate at the molecular level. *Nature*, 217(5129), 624–626.

Lenski, R.E. and Travisano, M. (1994). Dynamics of adaptation and diversification: A 10,000-generation experiment with bacterial populations. *Proceedings of the National Academy of Sciences*, 91(15), 6808–6814.

Monod, J. (1970). *Le hasard et la nécessité. Essai sur la philosophie naturelle de la biologie moderne*. Le Seuil, Paris.

Nattier, R., Pellens, R., Robillard, R., Jourdan, H., Legendre, F., Caesar, M., Nel, A., Grandcolas, P. (2017). Updating the phylogenetic dating of New Caledonian biodiversity with a meta-analysis of the available evidence. *Scientific Reports*, 7, 3705.

Rose, M.R. (1984). Artificial selection on a fitness-component in *Drosophila melanogaster*. *Evolution*, 7, 516–526.

14

Evolution in the Face of Chance

14.1. Introduction

In this chapter, we explain why the different sources of chance, which the living being is subjected to, can alter or even thwart the effectiveness of Darwinian selection and adaptation. Next we shall examine, through modeling in particular, the hypothesis according to which populations subjected to selection can under certain conditions actively respond to these sources of chance.

As a foundation, let us give an overly idealized representation of Darwinian evolution[1]:

– organisms with the same genotype (i.e. the same heritable information contained within their chromosomes) consequently have the same phenotype (i.e. the same physical characteristics, non-heritable);

– mutations generate heritable genotypic variation;

– natural selection increases the frequency of genotypes whose associated phenotype confers an advantage on the carrier in terms of *fitness* (i.e. the ability to leave descendants to the following generations: reproduction and survival).

As is well known, natural selection faces at least four sources of uncertainty that obscure this picture:

Chapter written by Amaury LAMBERT.

1 To make the link between microevolution, thus described, and macroevolution, that is to say the diversification of species, it should be added that the accumulation of genetic differences tends to prevent cross-breeding and promote speciation (i.e. to schematize: the separation of a species into two new species).

– *the stochasticity of phenotypic expression*, which we will call phenotypic noise or developmental noise, and which is the indeterminacy of the relationship between the genotype of an organism and its phenotype. Certain phenotypes of organisms having the same genome and living in identical environments can in fact be appreciably or even very different. Monozygotic twins ("identical twins") are often the same size, although this depends on how accurately they are measured. The under-determination of phenotype by genotype appears more clearly if we think of phenotypes such as age at puberty, age at first child and at menopause, the propensity to develop certain common diseases such as diabetes or cancer. However, the greater the expression noise, the less natural selection, which has influence only on the phenotypic expression of genes and not on the genotypes themselves, is able to filter, to retain the genotypes responsible for fitter future individuals;

– *randomness of mutations*: statistical regularities have been demonstrated in the occurrences of mutations on genomes as a function of target motifs or of certain mutagenic factors (see Alexandrov *et al.* 2013)[2], but ignorance of the contingent details of the process of germ cell division leads us to consider the mutations that occur in the germ line as random. Above all, it has been empirically verified innumerable times (starting with the Nobel Prize winners Luria and Delbrück (1943)[3]) that the mutational processes are not assisted by an environmental detection mechanism that would allow them to offer better adapted genotypes with greater certainty: mutations are fitness agnostic[4];

– *environmental variations*:   the biotic and abiotic characteristics of the environment vary in time and space, in a way which may be periodic or at least predictable (circadian cycles, tides) but is most often unpredictable for living organisms. Now, different media can promote different phenotypes. It is, for example, well-known empirically that cold environments favor endothermic organisms with a reduced surface area to volume ratio, resulting in a larger body and shorter extremities (as evidenced, for example, by the contrast between the arctic fox, heavy, stocky and small-legged, and the desert fox, small but slender with long legs and long ears). These responses of natural selection to the environment are generally gradual and cannot take place if the fluctuations of the environment over time are too rapid.

---

2 Alexandrov, L.B., *et al.* (2013). Signatures of mutational processes in human cancer. *Nature*, 500, 415–421.

3 Luria, S.E. and Delbrück, M. (1943). Mutations of bacteria from virus sensitivity to virus resistance. *Genetics*, 28(6), 491–511.

4 With the possible exception of certain Lamarckian inheritance mechanisms present in the immune system of prokaryotes, see, for example, Horvath and Barrangou (2010) and Müller *et al.* (2018).

There are of course evolved organisms that have developed mechanisms of phenotypic plasticity, that is to say, allowing for the modification of their phenotype in response to variations in the environment. But even in this case, these changes are not transmitted[5];

– *genetic drift*, that is to say, the chance of births and deaths naturally causing the frequencies of different alleles to fluctuate at each genetic locus in a population, independently from selection pressures. This genetic drift is all the stronger; the smaller the population is, the closer alleles are to neutrality, that is to say, only conferring a small advantage or disadvantage to their carrier from the point of view of fitness.

To summarize, we can think that the course of evolution is hampered or slowed down by chance, in the sense that chance imposes a certain degree of forcing to adaptation:

– insofar as mutations are agnostic to fitness, they suggest directions for evolution that can sometimes go in the direction of, but most often against, adaptation. We also know that most mutations are deleterious, that is to say, deteriorate the fitness of their carrier. The reason for this is not unique to living beings as we can understand from the well-known metaphor of the broken down car: if one were to randomly strike a hammer under the hood, there is a greater chance of damaging the engine than of repairing it;

– by reducing the efficiency of selection, phenotypic noise and genetic drift make the fixation of deleterious mutations more likely and that of beneficial mutations less likely;

– during a change of environment, if the optimal value of the phenotype under selection, or worse the very identity of the phenotype under selection, is different in the new environment as compared to the old environment, the adaptation must start from zero.

I would like to discuss here the idea that if the course of evolution is initially slowed down, or even hampered by chance, there are also *positive selection mechanisms as an active response to phenomena of a probabilistic nature*. To this end, we will examine three sources of chance (phenotypic noise, randomness of

---

5 With the exception of the heredity of certain acquired characters, in particular those mentioned in the previous note but not only those: cultural heredity, cytoplasmic heredity, etc., see Jablonka *et al.* (1998).

mutations, environmental variations)[6] and examine the reactions that selection can generate in response to these sources of chance.

I will then propose a simple model to illustrate these speculations. This model has no pretension in terms of realism for the biology that it wishes to embody (although it has its most generic properties) but has two characteristics which, as we will see, are sufficient (and undoubtedly necessary) conditions to allow evolution to respond actively to chance:

– the redundancy of phenotypic coding[7], that is to say, the fact that the same average phenotype can be associated with several different genotypes (and which is absolutely not incompatible with the fact that to one genotype and its average phenotype correspond several possible expressed phenotypes);

– the regularity of the statistical properties of chance, that is to say, the hypothesis that everything happens as if the sources of chance considered were produced through a probabilistic model whose parameters are constant over time.

14.2. Waddington and the concept of canalization

Let us now return, specifically, to the noise of phenotypic expression, that is to say the indeterminacy of the genotype-to-phenotype map. We use the term "noise" because this indeterminacy is usually only fairly relative. If the "genetic program" of the hand or the wing can indeed produce, even with a fixed genotype, limbs of slightly different sizes and colors, it always produces the same number of bones, the same number of joints and the same structure. Struck by the reliability and repeatability of this program, Waddington (1942) proposed the concept of *canalization*[8] in 1942: "The main thesis is that developmental reactions, *as they occur in organisms submitted to natural selection*, are in general canalized. That is to say, they are adjusted so as to bring about one definite end-result regardless of minor variations in conditions during the course of the reaction." Waddington's drawing reproduced in Figure 14.1 is a metaphor for the process of development or embryogenesis that starts from an initial result, the zygote or the bead at the top of an inclined surface, and leads to a single end result among several possible results. Here,

---

6 These are the four sources of chance listed above with the exception of genetic drift. One way to improve the efficiency of selection in response to genetic drift, for example, would be to increase the size of the population. However, it is difficult if not impossible, from an empirical as well as a theoretical point of view, to distinguish whether such an increase is the active result of selection or an effect derived from selection for organisms with a greater fitness giving the population a higher growth rate.

7 The idea of coding redundancy and its evolutionary implications is, for example, addressed by Whitacre and Bender (2010).

8 Waddington C.H. (1942). Canalization of development and the inheritance of acquired characters. *Nature*, 150, 563–565.

the important point is not, as the impression one can easily get at first glance, that several final positions are possible: the multiplicity of final positions is to be paralleled with the multiplicity of environments and therefore with the multiplicity of populations and species, or, at the level of the organism, with the multiplicity of cell types; but that for a given species and a given environment, only one final result is possible, and the point here is that the position of this final result is known the more precisely the path that leads there is channeled: canalized.

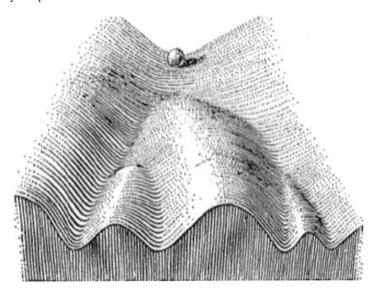

Figure 14.1. *The "epigenetic" landscape:*
a metaphor for development (Waddington 1957)

Figure 14.2 shows how the action of the genome, symbolized by the tutors which support the landscape (said at the time to be "epigenetic", but which today go by the less ambiguous name "embryogenetic"), controls and modifies the process of development. Waddington's idea is that selection favors canalized landscapes, that is to say, landscapes buffered against disturbances that occur during development.

A second idea from Waddington is that of channeling the expression of a phenotype toward an optimal end result in a given environment; the selection of a "pre-channeled" phenotypic expression toward an alternate optimal end result within an alternative environment; or even a "co-channeled" expression of other phenotypes.

A brilliant article written by Loison (2019)[9] on this subject demonstrates that this idea is based on the hypothesis that the considered species are supposed to have sufficiently evolved so as to possess mechanisms of phenotypic plasticity, that is to say of adaptive modification of the phenotype during life. In Figure 14.1, the apparent sensitivity of the paths to the initial conditions can also be interpreted as a representation of phenotypic plasticity. In the event of a definitive change of environment, the alternative path usually used by plasticity is recycled and reshaped through selection so as to become the only definitive, and this time irreversible, path taken by development. Waddington christened this phenomenon: "genetic assimilation".

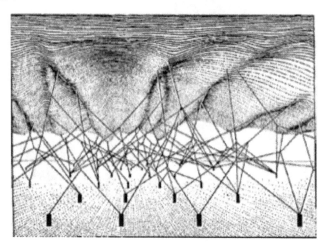

Figure 14.2. *The action of the genotype on the phenotype: view of the landscape from below shown in Figure 14.1 (Waddington 1957)*

Starting from Waddington's theories, whose metaphorical power sins a little by the absence of quantitative proof, we propose here to examine through the lens of modeling, the minimal conditions (in particular, those not requiring the existence of mechanisms for phenotypic plasticity) under which populations subjected to natural selection can react to the destabilizing action of chance. We have just mentioned the channeling, which is a reaction to the *noise of phenotypic expression*. We might also ask ourselves if the *effects of mutations* are not also shaped by evolution. As Waddington framed it in 1957, again, in the context of canalization:

9 Loison, L. (2019). Canalization and genetic assimilation: Reassessing the radicality of the Waddingtonian concept of inheritance of acquired characters. *Seminars in Cell & Developmental Biology*, 88, 4–13.

We have been led to conclude that natural selection for the ability to develop adaptively in relation to the environment will build up an epigenetic landscape which in its turn guides the phenotypic effects of the mutations available. In the light of this, the conventional statement that the raw materials of evolution are provided by random mutation appears hollow. The changes which occur in the nucleoproteins of the chromosomes may well be indeterminate, but the phenotypic effects of the alleles which have not yet been utilized in evolution cannot adequately be characterized as "random": they are conditioned by the modelling of the epigenetic landscape into a form which favors those paths of development which lead to end-states adapted to the environment.

Finally, we will study the way in which populations react to *environmental changes* occurring on a regular basis by proposing an elementary mechanism for storing the memory of these changes in the genome.

We now propose three possible selection responses to each of the three sources of uncertainty cited (noise of phenotypic expression, effects of mutations, environmental changes), we give them a name and try to illustrate them with a metaphor belonging to the field of decision-making:

– *canalization* favors genotypes that have the capacity, called "robustness", to produce the same phenotype despite the stochasticity of expression or development. Channeling consists of minimizing variance, like choosing to *cycle over taking the bus*: the best mode of transport to get to work is the one that gives us the assurance of a reliable travel time, protecting us against the vagaries of weather, waiting times and traffic jams;

– mutational *target selection* favors genotypes that are specific targets for future mutations. In the *design of a car*, we make inaccessible the parts that only a mechanic can repair but we facilitate access to parts more easily modified by the driver, such as the fuel tank, the oil pan and the radiator;

– *neighborhood selection*[10] favors genotypes accessible through mutation to genotypes adapted to alternative environments. In the underground stations of a large city, there are always *two exits*, one at the front end and one at the back end of the train, which one usually takes interchangeably. However, if you note that in the majority of cases the newsagent is located at the front exit, you may become accustomed to always exiting that way: between two choices that are generally indifferent, in the case you wish to buy a newspaper, you make the choice to take the shorter route.

---

10 The expression is taken from Olivier Rivoire (CNRS, Collège de France).

14.3. A stochastic model of Darwinian evolution

14.3.1. *Redundancy and neutral networks*

The sequence of a messenger RNA is a series of letters taken from the alphabet $\{A, C, G, U\}$. It can also be seen as a succession of three-letter words, called *codons*. The protein translation of this sequence is done according to a universal code that makes each codon correspond to a unique amino acid, among the 20 of which, all proteins are made. If we remove from the list of $4 \times 4 \times 4 = 64$ possible codons, the three stop codons UAA, UAG and UGA that signal the stop of translation, we see that the genetic code matches 20 amino acids to 61 possible codons. There are therefore different codons that code for the same amino acid, as is the case, for example, of the four codons starting with CG, which all code for arginine, or the GAA and GAG codons which both code for glutamate. These codons are said to be *degenerate* and the genetic code is *redundant*. In mathematics, we say that the relation that maps the genotype (the initial DNA sequence or its transcribed messenger RNA) to its phenotype (the translated protein of the messenger RNA) is not injective (i.e., not "one-to-one"). Another strongly non-injective relationship is that which maps an amino acid or protein to its function in the body.

Let us represent the space of all possible genotypes as a graph or network of which each vertex is a sequence. Each pair of sequences is connected by an edge in this network provided there is a way to switch from one to the other through a single *substitution* (i.e. the replacement of one nucleotide base by another). In this representation, two neighboring vertices therefore differ at a single site of their sequences. If the codons supporting this site encode the same amino acid, both sequences encode the same protein. Taking into account the redundancy of the genetic code, it is therefore possible to travel in genotype space by borrowing the edges of the graph but without changing the corresponding phenotype.

We say that a subgraph of genotypes is a *neutral network* if it is a connected set in which all the elements have the same phenotype, or more simply, the same levels of fitness. Examples of neutral networks in biology abound (metabolic pathways performing the same function, proteins binding to the same substrate), and it is often possible to travel from one end of the global network to the other without straying too far from one such subgraph (see Wagner 2008)[11].

In the following section, we choose to model the genotype by a series of characters with values in an alphabet with three symbols (instead of four symbols for the DNA

---

11 Wagner, A. (2008). Neutralism and selectionism: a network-based reconciliation. *Nature Reviews Genetics*, 9(12), 965.

and RNA sequences, and 20 symbols for the proteins) and propose several quantitative phenotypes associated redundantly (i.e., in a non-injective way) to the genotype.

14.3.2. A toy model

From here on, we shall seek to implement our ideas about canalization, target selection and neighborhood selection in a stochastic model of Darwinian evolution. We will term a *genotype* as a sequence $\varepsilon = (\varepsilon_1, \ldots, \varepsilon_L)$ of L characters taken from the $\{-1, 0, +1\}$ alphabet, which we will sometimes refer to as "spins", by analogy with but unrelated to physics (electromagnetism, quantum mechanics), where the spin is a particle property (taking only in physics the value -1 or the value $+1$).

The *expected* or *average phenotype* $X(\varepsilon)$ associated with the genotype ε is equal to the sum of all the spins of the genotype:

$$X(\varepsilon) = \sum_{i=1}^{L} \varepsilon_i$$

This type of model is common in quantitative genetics, where the value of a quantitative trait associated with a given genotype is the sum of the so-called allelic effects at each locus controlling this trait. Here, the allelic effects can only take the values -1, 0 and $+1$.

In addition to its simplicity, this model allows one to see the phenotype as the result of a process: each spin corresponds to an instruction, for example, to take a step to the right for $+1$, to take a step to the left for -1, to stay put for 0; and the phenotype is the result obtained after following the list of L instructions.

To model the noise of phenotypic expression, we will assume that the *expressed phenotype* $\tilde{X}(\varepsilon)$ associated with the genotype ε is defined by:

$$\tilde{X}(\varepsilon) = \sum_{i=1}^{L} a_i \varepsilon_i$$

where the (a_i) are positive random variables, which all have an expectation equal to 1. We will suppose that the *multiplicative noise* a_i due to the site i is of variance σ_i, which may depend on i; possibly zero if the i site does not contribute to the phenotypic noise. We will assume that these variances are fixed and identical for all individuals, but that their realizations are independent between individuals.

In the following, we will assume that the *fitness* of an individual of expressed phenotype x is $\phi(x)$, where ϕ is a positive map that reaches its maximum at 0, that is to say, the optimal value of the phenotype is 0. In this case, we say that the selection is stabilizing, because it tends to make the average for the phenotypes of the population return to a central value. In the simulations, we will take:

$$\phi(x) = \exp(-cx^2/2)$$

where $c > 0$ is called selection intensity: the smaller c, the more the fitness map is spread out and therefore not very sensitive to phenotypic differences.

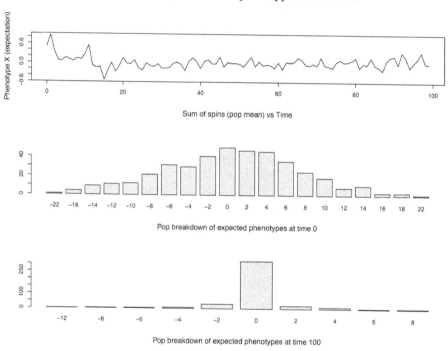

Figure 14.3. *Simulation of the evolution of a population of $N = 300$ individuals over 100 generations. Each individual is endowed with a genotype of length $L = 50$ spins equal to -1 or $+1$. The associated expected phenotype X is the sum of the spins of the genotype and the expressed phenotype $\tilde{X}$ is the noisy sum (see text), whose fitness is maximum at 0. (Top) Evolution of the average expected phenotype over time. (Middle) Distribution of expected phenotypes at time $t = 0$. (Bottom) Distribution of expected phenotypes at time $t = 100$. Numerical values: $c = 2.5$, $\mu = 0.1$, $p_{-,+} = p_{+,-} = 0.02$*

14.3.3. *Mutation-selection algorithm*

We explain here the way in which we are going to model the evolution over time for a population subjected to natural selection, reminiscent of the so-called Wright–Fisher model (see Lambert 2008)[12].

We give ourselves a population of constant size equal to N. Each individual $j \in \{1, \ldots, N\}$ in this population has a genotype $\varepsilon(j)$ as described in the previous section and a fitness $\varphi(j)$ calculated by one of the methods described in the previous section. At each generation t, the population is renewed in the following manner. Each of the individuals of generation $t + 1$, independently, has a probability of being the child of individual j of generation t, and therefore of inheriting the genotype $\varepsilon(j)$, proportional to $\varphi(j)$.

Once the population is completely renewed, each individual of generation $t + 1$ independently sees its genotype mutate with the same probability μ. To generate a mutant genotype from a target genotype, each spin of its sequence mutates independently according to transition probabilities (p_{ij}) (where i and j are elements of $\{-1, 0, +1\}$ and p_{ij} denotes the probability that the spin i mutates into the spin j) that we will specify.

Finally, in each of the three subsections, the initial distribution of the N genotypes of the population is assumed to be given by the stationary distribution of the mutational process, as if the population had previously evolved for a long time in a neutral manner, that is to say, in the absence of selection. An example of simulation is shown in Figure 14.3. Note the concentration of phenotypes around the optimal fitness value at time $t = 100$.

14.4. Numerical results

14.4.1. *Canalization*

We will assume in this section that the spins can only take the values -1 or $+1$ and that the sum of the spins giving the phenotype is only taken from a subset S of $\{1, \ldots, L\}$ called the *coding part*, whose identity is inherited with the sequence and is also mutable. In other words, it is considered that the expressed phenotype is now:

$$\tilde{X}(\varepsilon) = \sum_{i \in S} a_i \varepsilon_i$$

---

12 Lambert, A. (2008). Population dynamics and random genealogies. *Stochastic Models*, 24(1), 45–163.

The data of the sequence $\varepsilon = (\varepsilon_1, \ldots, \varepsilon_L)$ and the data of the set S of coding sites are therefore assumed here to be transmitted to the progeny. When the S set mutates, it is equally probable to win or lose a site.

In the absence of selection, the cumulative result of mutations over time is to cause the set S to lose on average only one or two sites (a reflected symmetric random walk). In the presence of selection, we see in Figure 14.4 that the coding part loses 20 sites in 1,000 generations (bottom panel), which results in a drastic reduction of the phenotypic noise (middle panel). Note that the coding sites are apparently lost 2 by 2, since phenotypes corresponding to odd cardinal coding sets cannot take the maximum fitness value of 0 (the value 0 for spins is prohibited in this section).

Here, the variances σ_i of the multiplicative noises are fixed *a priori* but vary along the sequence. Sites that contribute significantly to phenotypic noise are at first masked, as they frequently produce suboptimal phenotypes eliminated by selection. The sites contributing very little to the noise are not detected by selection so that the coding part ends up reaching a minimum size, below which it can only pass by chance. Note also that if the variances σ_i of multiplicative noises are initially too high, selection cannot distinguish between expected suboptimal phenotypes and very noisy phenotypes, which reduces both the efficiency of the selection of good genotypes ($X(\varepsilon)$ close to 0) and the efficiency of canalization, that is to say selection for the genotypes generating less noise (small S). These results are perfectly illustrated in the sentence by Waddington quoted above: "[Development is] adjusted so as to bring about one definite end-result regardless of minor variations in conditions during the course of the reaction".

Canalization has two other consequences, both of which relate to the ability to evolve in the future (sometimes referred to as "evolvability"). First, if the sites contributing to the expression noise of the X phenotype contribute in the same way to the expression noise of other phenotypes, masking these sites allows these other phenotypes to be canalized simultaneously, even if they are not (yet) under selection (co-channeling). Next, masking sites temporarily allows genetic variation to silently build up and therefore produce diversity without diminishing fitness. This diversity can be used as a raw material for selection in subsequent situations of environmental stress[13]. Waddington (1957), again, wrote in 1957[14]: "It is advantageous to a population to contain some genic variability to cope with environmental changes and to give the potentiality for evolutionary advance. The more strongly the epigenetic system is buffered against genic variation, the more reserve of such variability the population will be able to contain without endangering the attainment of the optimum in the normal environment. There will thus be a selective pressure in favour

---

13 See the example of the Hsp90 chaperone protein that would buffer the effects of mutations in Rutherford and Lindquist (1998), but also see, for a review, Geiler-Samerotte *et al.* (2016)

14 Waddington C.H. (1957). *The Strategy of the Genes*. Allen and Unwin, London.

of epigenetic systems which can absorb some genic variation without this producing any phenotypic effects, although, as we have seen, this lack of response to variation of genes cannot be pushed too far".

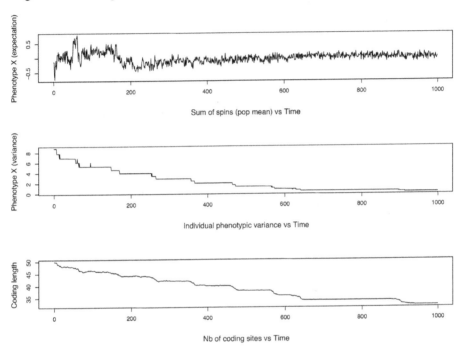

Figure 14.4. *Canalization: Simulation of the evolution of a population of $N = 400$ individuals during 1,000 generations. Each individual is endowed with a genotype of length $L = 50$ spins equal to -1 or $+1$. The associated expected phenotype X is the sum of the spins of the coding part S of the genotype, which is mutable and heritable, and the expressed phenotype is the noisy sum $\tilde{X}$ (see text), whose fitness is maximum at 0. (Top) Evolution of the average expected phenotype over time. (Middle) Evolution of phenotypic variance over time. (Bottom) Cardinal of the coding part over time. Numerical values: $c = 2.5$, $\mu = 0.1$, $p_{-,+} = p_{+,-} = 0.02$, mutation probability of $S = 0.06$, $\sigma_i \approx 0.3$ (varying along the sequence)*

Note that some authors have argued that canalization could also, in some species, be simply a by-product of the complexity of developmental processes[15].

15 Siegal, M. L. and Bergman, A. (2002). Waddington's canalization revisited: Developmental stability and evolution. *Proceedings of the National Academy of Sciences*, 99(16), 10528–10532.

14.4.2. *Target selection*

In this section, we return to the initial framework where all the sites are coding, the phenotype has an optimal value equal to 0 and the selection is made on the basis of the described phenotype, which is the sum of the ε_i spins scaled by a multiplicative noise variable a_i.

This time, all three spin values are possible. We also assume the existence of a *mutational bias*, since $p_{0,-} = p_{0,+}$ but $p_{-,0} > p_{+,0}$. In the absence of selection, at mutational equilibrium, the sequences therefore contain more $+1$ than -1 and the average X phenotype takes positive values very far from its optimal value 0 (see Figure 14.5).

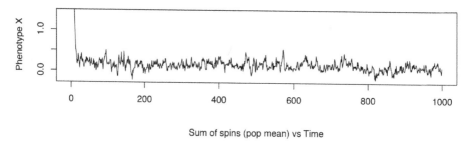

Sum of spins (pop mean) vs Time

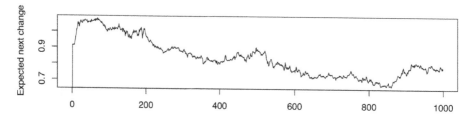

Expected phenotypic change (pop mean) vs Time

Figure 14.5. *Target selection. Simulation of the evolution of a population of $N = 200$ individuals during 1 000 generations. Each individual is endowed with a genotype of length $L = 50$ spins equal to -1, 0 or $+1$. The associated expected phenotype X is the sum of the spins of the genotype and the expressed phenotype $\hat{X}$ is the noisy sum (see text), whose fitness is maximum at 0. (Top) Evolution of the average expected phenotype over the course of time. (Bottom) The expected phenotypic effect of the next mutant as a function of time. Numerical values: $c = 2.5$, $\sigma_i \approx 0.15$, $\mu = 0.1$, $p_{0,-} = p_{0,+} = 0.04$ and $p_{-,0} = 0.08$, $p_{+,0} = 0.02$ (mutational bias in favor of positive spins)*

We are interested here in the phenotypic effect of mutations and its evolution under the effect of selection. Assuming we know the state of the population at time t, we denote by ΔX_t the change in phenotype caused in the first mutant immediately after time t. It is not difficult to see that at mutational equilibrium this phenotypic effect is zero in terms of expectation, that is to say, $\mathbb{E}(\Delta X_0) = 0$ (as it is difficult to see in Figure 14.5 because the frame is cut to better visualize the rest of the process). As selection shifts the distribution of genotypes out of mutational equilibrium, $\mathbb{E}(\Delta X_t)$ increases sharply within a few generations due to the excess of -1 due to selection, compared to the equilibrium state. In the rest of the process, the population remains predominantly composed of zero-sum genotypes, but the proportion of zeros within the spins of these genotypes increases. Since the zeros are not sensitive to multiplicative phenotypic noise, their presence reduces the phenotypic variance without modifying the expected value of the phenotype. The more zeros there are in the sequences, the closer $\mathbb{E}(\Delta X_t)$ is to 0, its initial value, because the effect of a mutation on a zero spin is zero on average ($p_{0,-} = p_{0,+}$). This phenomenon is very well illustrated by Waddington: "Phenotypic effects of mutations] are conditioned by the modelling of the epigenetic landscape into a form which favours those paths of development which lead to end-states adapted to the environment".

On the other hand, it would be risky to draw a general conclusion as to the supposedly beneficial consequences of target selection on future development. The variations in the phenotypic effects of the mutations in this example are only a by-product of selection.

14.4.3. *Neighborhood selection*

In this section, we assume the alternation of two environments, denoted 0 and 1. In environment 0, the fitness of an individual of genotype ε is given by $\phi(X(\varepsilon))$ (without the addition of phenotypic noise, to focus on the effect of environmental variations). In environment 1, the fitness of an individual of genotype ε grows with the value of an alternative phenotype $Y(\varepsilon)$ chosen among three possible phenotypes, which we will now describe. We will need to introduce the notation $\mathbb{1}_A$, called the indicator function of the event A, which is equal to 1 if A occurs, and 0 if otherwise.

The first alternative phenotype $Y_1(\varepsilon)$ is defined as:

$$Y_1(\varepsilon) = \sum_{i=1}^{L-1} \mathbb{1}_{\{\varepsilon_i \varepsilon_{i+1} = -1\}}$$

In other words, $Y_1(\varepsilon)$ counts the number of times in the sequence ε where the product of two consecutive spins is worth -1, that is, the number of *sticky pairs*, where a sticky pair is a pair of neighboring spins of type $(+1, -1)$ or $(-1, +1)$.

The second alternative phenotype $Y_2(\varepsilon)$ is defined as:

$$Y_2(\varepsilon) = \max_{j\in\{-1,0,+1\}} \left(\sum_{i=1}^{L} \mathbb{1}_{\{\varepsilon_i=j\}}\right)$$

that is to say, $Y_2(\varepsilon)$ counts the number of occurrences in the sequence ε of the most frequent spin in that sequence. Finally, the third alternative phenotype $Y_3(\varepsilon)$ is defined as the longest constant subsequence of ε:

$$Y_3(\varepsilon) = \max\{b - a + 1 : 1 \le a \le b \le L, \varepsilon_a = \cdots = \varepsilon_b\}$$

Figures 14.6 and 14.7 present two simulations (independent and using the same parameters) where the environmental variations are more or less periodic, in the case where the alternative phenotype is Y_1. Just before visiting environment 1, and because of the selection at work in environment 0, the adapted genotypes carry roughly the same number of -1 and $+1$, distributed evenly over the sequence. Thus, a constant subsequence of a suitable genotype is typically of length 2, which results in a number of sticky pairs of the order of $L/2$. As can be seen in the figures (bottom panel), genotypes with a greater number of sticky pairs increase in frequency in environment 1. Back in environment 0, selection restores balance between positive and negative spins but by slowly modifying their distribution over the sequence, partially preserving the number of sticky pairs: neighborhood selection favors genotypes of the "zero sum" neutral network that are close to the spatial regions of genotypes that have a large number of sticky pairs. This memory of environment 1 can facilitate future adaptation for the next return of environment 1, as can be seen particularly in Figure 14.6.

We carried out the same simulations with the alternative phenotypes Y_2 and Y_3 and obtained the same results with Y_2 but not with Y_3. In the case of the latter, the selection in environment 1 favors long constant subsequences, but back in environment 0, the memory of this selection is very quickly erased by the mutations, which most likely fall first inside of these long subsequences. In the case of Y_2, the first occurrences of environment 1 see the genotypes with a large number of $+1$ or a large number of -1 increase in frequency. Back in environment 0, the memory of this selection is very quickly erased by the re-establishment of the balance between positive and negative spins. However, neighborhood selection is effective as soon as the selection in environment 1 causes the genotypes with a large number of zeros to increase in frequency, the memory of which is retained when returning to environment 0.

Therefore, neighborhood selection requires well-connected neutral networks: the zero sum network is well connected to the networks of genotypes that have a large number of sticky pairs and to the network of genotypes that have a large number of zeros, but not to the network of genotypes containing a long, constant subsequence.

Likewise, neighborhood selection requires sojourn times in each environment long enough to allow time for adaptation to take place, but short enough to prevent genetic drift from erasing the memory of the selection that operated in the previous environment.

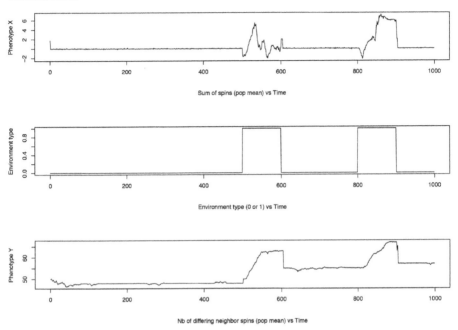

Sum of spins (pop mean) vs Time

Environment type (0 or 1) vs Time

Nb of differing neighbor spins (pop mean) vs Time

Figure 14.6. *Neighborhood selection. Simulation of the evolution of a population of $N = 200$ individuals during 1,000 generations in a variable environment. Each individual is endowed with a genotype of length $L = 100$ spins equal to -1, 0 or $+1$. The associated phenotype X is the sum of the spins of the genotype, and the Y phenotype is the number of sticky pairs (see text). In environment 0, fitness is maximized when X is close to 0; in environment 1, it grows with Y. (Top) Evolution of the average X phenotype over time. (Middle) Changes in the environment over time. (Bottom) Evolution of the average Y phenotype over time. Numerical values: $c = 2.5$, $\mu = 0.01$*

Under these assumptions (connectivity of neutral networks corresponding to optimal phenotypes in each environment, periodicity of environments in the correct timescale), neighborhood selection increases the ability to adapt to future environments, provided they have already been visited.

The results of the various simulations exhibited here will be discussed in the next section.

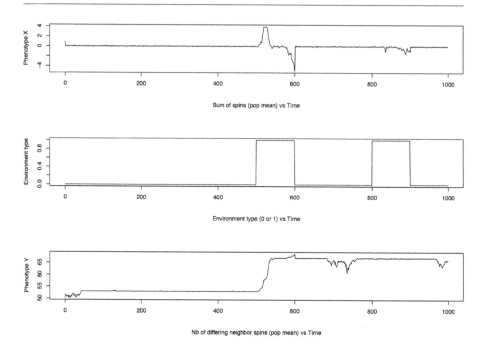

Figure 14.7. *Neighborhood selection. Simulation of the evolution of a population under the same assumptions as in Figure 14.6*

14.5. Discussion

The subject of this communication was incidentally a pretext to demonstrate the effectiveness of mathematical models and stochastic simulations in helping to understand phenomena that are *not* or *hardly repeatable*, such as evolutionary processes, which have the double disadvantage of being very slow on a human scale and of inextricably presenting contingent and selective aspects[16].

The second point, addressed at the start of the chapter, is the idea that the efficiency of adaptation is intrinsically reduced by four sources of chance, which blur its ability to filter the genotypes corresponding to the best adapted phenotypes: the stochasticity of phenotypic expression, the randomness of mutations, environmental variations and genetic drift.

The third more subtle point is that populations can react by *evolving in directions that are not entirely blind or contingent, in the face of these sources of chance.*

---

16 With the exception of evolutionary experiments that can be carried out on organisms with a short generation time such as bacteria, yeasts or insects, or even directly on molecules (proteins): these experiments are relatively rapid and their conditions are controlled.

Phenotypic noise is reduced by canalization, the effects of mutations are altered by target selection, and neighborhood selection moves genotypes into regions of space at the intersection of, to put it bluntly, sets of genotypes suitable to one of the environments.

In particular, canalization and neighborhood selection *actively promote the ability to cope with an uncertain environment*:

– provided that the environment changes in timescales that are both long in the scale of mutations (to allow for adaptation) and short in the scale of genetic drift (to prevent memory erasure by neutral evolution);

– and provided that it is sufficiently ergodic, that is to say, regular in its statistical properties: in our model, the contribution to the phenotypic noise of each site is different at each generation but their mean and variance remain constant over time; the succession of environments occurs periodically (it could also be Markovian).

In the three cases, we were able to illustrate these *evolutionary reactions in the face of chance*, under the hypothesis that the genotypes are quite strongly redundant: to one (expected) phenotype can correspond multiple different genotypes. We can calculate exactly the (large) number of sequences of spins with values in $\{-1, 0, +1\}$ whose sum X takes a fixed value (the same for the phenotypes Y_1, Y_2 and Y_3), and this number is moreover maximum for the optimal value 0. In the case of canalization, the number of coding sites can change without changing the expected phenotype; in the case of target selection, the number of zeros in the sequence can change without changing the expected phenotype; in the case of neighborhood selection, the distribution of spins in the sequence can also change without changing the phenotype. Under this hypothesis of redundancy, the genotypes associated with the same expected phenotype form neutral networks, that is to say connected subgraphs (in the sense of mutations) whose elements have the same fitness. The *memory of past disturbances* is encoded in genomes through the selection of specific regions of these neutral networks.

Several wide-open questions arise from these speculations, which are as follows: to what extent these mechanisms are generic; whether their effects are beneficial in the long term (i.e. whether they generally promote future adaptation); and whether the memory of disturbances can be encoded in genomes or elsewhere in a harder and therefore more durable way.

14.6. Acknowledgments

The author would like to thank François Blanquart, Emmanuel Schertzer and especially Olivier Rivoire for their proofreading.

15

Chance, Contingency and the Origins of Life: Some Historical Issues

Was the origin of life on Earth an accidental event in which chance played a key role? An intuitive answer to this issue would try to find an equidistant point in the complex interplay between purely random events and the narrow constraints defined by the laws of physics and chemistry. Current descriptions of the appearance of living systems have moved from the simplistic assumption that life emerged directly from a primordial soup into a set of more refined hypotheses suggesting that the origin of living systems was the outcome of a historical process during which the molecular structures and functions of cells originated and evolved. Of course, acceptance of the hypothesis that the first living entities were the outcome of a lengthy (but not necessarily slow) process of prebiological evolution does not evade the issue of the ontological role of randomness and historical contingency in defining the properties of life. It is not easy to find acceptable answers to these questions. Our understanding of the origin of living systems is hindered, among other issues, by the lack of geological evidence of the environment in which they arose, the absence of fossil records of prebiological systems that preceded the first cells and, equally significant, the lack of an all-embracing, generally agreed definition of what life is.

Given the many unknowns that plague our understanding of the origin of life, it is difficult to make a proper assessment of the role of contingency in the succession of events that led to the universal biochemical traits and the exquisite molecular interactions that characterize all cells. Attempts to understand these issues have far-reaching consequences on controversies about the likelihood of extraterrestrial life, as well as controversies about the development of experimental and theoretical

Chapter written by Antonio Lazcano.

research programs on the origins of life, synthetic biology and, ultimately, on a proper understanding of the essential traits of living systems.

*

After the publication of *On the Origin of Species* in 1859, many naturalists adopted the idea that living organisms were the historical outcome of gradual transformation of lifeless matter, although Darwin did not discuss the origin of life and even appealed to the terms "creation" and "Creator" to refer to it. His inner feelings are revealed, however, in a letter sent in 1863 to his friend J.D. Hooker, to whom he wrote that "[…] I have long regretted that I truckled to public opinion and used the Pentateuchal term of creation, by which I really meant 'appeared' by some wholly unknown process" (Peretó *et al.* 2009).

Darwin did include few statements on the emergence of life in published works, but emphasized over and over again the lack of evidence on the likelihood of spontaneous generation. This was quite disturbing for some of his followers, including the German geologist Heinrich George Bronn and the naturalist Ernst Haeckel. As the latter wrote, "... the chief defect of the Darwinian theory is that it throws no light on the origin of the primitive organism, probably a simple cell, from which all the others have descended. When Darwin assumes a special creative act for this first species, he is not consistent, and, I think, not quite sincere" (see Peretó *et al.* 2009).

Haeckel was firmly convinced that since there was no essential difference between the living and inert matter and that natural processes stand in a material, causal and historical connection, a direct link can be established between them. Based on his Monistic idea of the unity of nature, Haeckel proposed a somewhat teleological scheme of evolution that included the spontaneous origin of life. He had a major influence in the ideas of Oparin, who nevertheless rejected the possibility of spontaneous emergence of life and who, in 1924, proposed the available chemical data within an evolutionary sequence to argue that the first organisms had been preceded by a stage of prebiotic evolution during which the abiotic synthesis of organic compounds takes place (Lazcano 2016a).

By the turn of the 20th century, the view that the microscopic world could be described statistically, assuming random atomic movements, had found a firm place in biochemistry, where many saw enzymes as colloidal catalysts. This led to suggestions that entities smaller and simpler than protoplasm itself could be alive (see Fry 2006). In 1917, Felix d'Herelle discovered a self-propagating filterable "substance" that attacked and dissolved bacilli, which were later identified as viruses. Viruses are small and simple, and in the eyes of d'Herelle these traits evidenced their primitiveness, leading him to propose in his 1924 book *Immunity in Natural Infectious Diseases* a schema in which a hypothetical bacteriophage which he named *Protobios protobios* was proposed as the evolutionary ancestor of both plants and animals. As d'Herelle wrote, "the ultravirus, or Protobe, which represents the most rudimentary being possessing elementary life, is nevertheless endowed with the same 'powers' as being much higher in the scale of organization" (see Summers 1999).

Speculation on the origins of life was rife, and during the same epoch, the polymath Leonard Troland, who had been trained in Bolztmann's tradition of statistical physics, proposed that the first life form had been a self-replicating enzyme-like molecule formed spontaneously by random events in the primitive oceans. This primordial enzyme was assumed by Troland to be endowed with autocatalytic properties that allowed it to self-multiply, as well as having heterocatalytic abilities that could alter its surroundings, therefore giving rise to metabolism.

It did not take long for Hermann J. Muller, an American geneticist who had started to work on the role of X-rays in mutagenesis in the T.H. Morgan group, to adjust Troland's hypothesis and propose that the ancestral molecule had been, in fact, a primordial gene. In a paper published in 1922, Muller compared genes with viruses, and wrote that "[...] that two distinct kinds of substances – the d'Hérelle substances and the genes – should both possess this most remarkable property of heritable variation or 'mutability', each working by a totally different mechanism, is quite conceivable, considering the complexity of protoplasm, yet it would seem a curious coincidence indeed... It would be very rash to call these bodies genes, and yet at present we must confess that there is no distinction known between the genes and them" (Muller 1922).

In spite of their mutual animosity, Morgan and Muller shared their appreciation of genetic mutation as the fundamental mechanism of evolutionary novelties. Even more explicitly than Troland, Muller (1922, 1926) argued that the first living material was formed abruptly by purely random events and consisted of little more than a mutable gene, or a set of genes, endowed with catalytic and autoreplicative properties.

Muller held quite strongly to his reduccionist hypothesis, and following the *in vitro* enzymatic synthesis of strands of DNA reported by Arthur Kornberg and his associates (Lehman *et al.* 1958), during the University of Chicago 1959 celebrations of the centennial of Darwin's *On the Origin of Species*, stated that "those who define life as I do will admit that the most primitive forms of things that deserve to be called living had already been made in the test tube by A. Kornberg" (Muller 1960). The development in molecular genetics and the consolidation of prebiotic chemistry allowed Muller to update his gene-first proposal by arguing that what had emerged in the primitive oceans had been, in fact, a primordial DNA molecule: "[...] it is to be expected that at last, just before the appearance of life, the very ocean had become, in Haldane's vivid phraseology, a gigantic bowl of soup", wrote Muller, and added "drop into this a nucleotide chain and it should eventually breed!" (Muller 1960). A few years later Muller stated that

> [...] life as we know it, if stripped of all its superstructures, lies in the three faculties possessed by the gene material. These may be defined as, firstly, the self-specification, after its own pattern, of new material produced by it or under its guidance; secondly, of performing this operation even when it itself has undergone a great succession of permanent pattern changes which, taken in their totality, can be of a practically unlimited diversity; thirdly, of, through these changes, significantly and (for different cases) diversely affecting other materials and, therewith, its own success in genetic survival.

Muller (1961) added that "the gene material alone, of all natural materials, possesses these faculties, and it is therefore legitimate to call it living material, the present-day representative of the first life".

In other words, for Muller (and many others), the essence of life lies in the combination of autocatalysis, heterocatalysis and evolvability, which he reduced to chance events. Thus, in assuming the sudden appearance of a self-replicating gene or DNA molecule, independently of the evolution of the metabolic heterocatalytic pathways, Muller was in fact granting chance a key ontological role, and was assuming that life could be so well defined that the exact point at which it started could be established with the sudden appearance of the first DNA molecule. He was also stating his reductionism: DNA alone is functionally inept. DNA molecules are not alive, and replicate and express their information and evolve only when they are part of a cell or under *in vitro* conditions that simulate the intracellular environment. In fact, genetic information is embodied in cells, which are spatially and functionally limited chemical systems.

In his famous 1970 book *Chance and Necessity*, Jacques Monod stressed the significance of the molecular description of the genetic code that allowed the construction of a purely physical theory of heredity, which he described as the "secret of life", and added that "[...] it might be thought that the discovery of the universal mechanisms basic to the essential properties of living beings would have helped solve the problem of life's origins. As it turns out, these discoveries, by almost entirely transforming the question, have shown it to be even more difficult than it formerly appeared" (Monod 1970). He was referring, of course, not only to the double helix model of DNA, but also to his own work on the exquisite structure and functions of regulatory proteins (Morange 2012). The material nature of life could be understood by the physical and chemical properties of the structures, functions and interactions of nucleic acids and proteins, but at the time the question of how these molecules had appeared and how genetic polymers came to direct protein synthesis appeared to be unanswerable.

For Monod, the appearance of life was a highly improbable accident, a conclusion in odds with the heterotrophic theory proposed by Oparin. This scientific antagonism must be seen against the background of intense ideological issues and political tensions including Oparin's proximity with Stalin's regime and his support of Lysenko, which Monod strongly opposed. Their animosity can be explained not only by their philosophical and political differences, but also by the fact that at the time many molecular biologists assumed that the basic traits of cells could be explained by random mutations. As Mayr (1988) wrote "[...] although Monod is adamantly opposed to determinism, including the invoking of final causes, he totally ignores natural selection as a creative process and ascribes all evolution to pure chance". Although it is true that by the time Monod wrote his book proposals of an RNA world already had been made (Lazcano 2012), but an understanding of the relationship between proteins and nucleic acids was far from being understood. Monod's views would have far-reaching consequences, and would eventually lead Popper (1974) to write that the issue of the origin of life is "an impenetrable barrier to science and a residue to all attempts to reduce biology to chemistry and physics" (see Fry 2002).

How the ubiquitous nucleic acid-based genetic system of extant life originated remains one of the major unsolved problems in contemporary biology. The discovery of the catalytic properties of RNA has led to the hypothesis of an RNA world from which extant cells evolved. Regardless of the replicative, catalytic and regulatory properties of the highly ubiquitous RNA molecules, neither RNA nor any other substance is alive by itself. Life depends on individual molecule systems, their subcellular interactions and the functional integration between them and with the environment. These extremely complex and refined networks are a result of the evolution of the system as a whole. Cells are made up of individual components and systems of molecules that interact with each other. Many of these subcellular units have a modular structure and exhibit physical–chemical and biochemical properties that remain active even when isolated. Therefore, it should not surprise us that many of these components exhibit properties associated with life, such as catalysis, replication, mutation and self-assembly.

The biochemical properties of subcellular systems and their components, including nucleic acid replication, enzymatic catalysis and membrane forming properties, are the historical outcome of the evolutionary amplification of the basic physico-chemical properties that their simpler molecular predecessors also possessed. The structure and functional properties of the molecular components of the cells, together with the subcellular systems they form and their complex adaptive properties, can only be understood as the historical results of a combination of physical-chemical constraints and natural selection, which reflect the interplay of a great number of historical contingencies. For instance, although the amphiphilic nature of archaeal and bacterial lipids leads to their self-organization in liposomes and micelles due to purely physico-chemical reasons, their strikingly different chemical compositions are the evolutionary outcome of a historical process that took place during the early stages of cell evolution. In other words, there was no spontaneous origin of the exquisite structure and functional properties of nucleic acids, enzymes, or metabolism, but it is only through an extension of Darwinian evolution to the subcellular components that their emergence can be understood (Lazcano 2016).

Accordingly, to understand the nature of the living we must recognize both the limits imposed on living beings by the laws of physics and chemistry, as well as the contingent nature of evolutionary history. As Mayr (1997) wrote some time ago, "biology is, like physics and chemistry, a science. But biology is not a science like physics and chemistry; it is rather an autonomous science on a par with the equally autonomous physical sciences". One of the main differences between biology and these other disciplines is, of course, its historical nature. For example, concepts such as natural selection and genetic program are consistent with physical laws, but cannot be deduced from them. Although it is impossible to predict the contingent events that determine biological evolution, these are explainable in retrospective terms.

In other words, to understand the nature of life we must recognize both the limits imposed by physics and chemistry, as well as the role of history's contingency. As summarized elsewhere (Lazcano 2016b), prebiotic processes did not have a fully deterministic character, nor was the origin of life a continuous, unbroken chain of progressive chemical transformations steadily proceeding to the first living systems. We do not know whether or not the transition to life from non-living systems requires a rather narrow set of environmental constrains, and we cannot discount the possibility that even a slight modification of the primitive environment could have prevented the appearance of life on Earth. As the French philosopher Pascal once wrote, had Cleopatra's nose been different, the course of history may have changed and the same may be true of the historical development of life.

15.1. Acknowledgments

I am indebted to Drs. José Alberto Campillo Balderas and Ricardo Hernández Morales for their help with the manuscript.

15.2. References

Fry, I. (2002). *The Emergence of Life on Earth*. Rutgers University Press, New Brunswick.

Fry, I. (2006). The origins of research into the origins of life. *Endeavour*, 30. 24–28.

d'Herelle, F. (1924). *Immunity in Natural Infectious Diseases.* Williams and Wilkins, Baltimore, MD.

Lazcano, A. (2012). The biochemical roots of the RNA world: From zymonucleic acid to ribozymes. *History and Philosophy of the Life Sciences*, 34, 407–424.

Lazcano, A. (ed.) (2016a). Precellular evolution and the origin of life: Some notes on reductionism, complexity and historical contingency. In *What is life? On Earth and Beyond.* Cambridge University Press, Cambridge.

Lazcano, A. (2016b). Alexandr I. Oparin and the origin of life: A historical reassessment of the heterotrophic theory. *Journal of Molecular Evolution*, 83, 214–222.

Lehman, I.R., Zimmerman, S.B., Adler, J., Bessman, M.J., Simms, E.S., Kornberg, A. (1958). Enzymatic synthesis of deoxyribonucleotic acid V. Chemical composition of enzymatically synthesized deoxyribonucleic acid. *Proceedings of the National Academy of Sciences of the USA*, 44, 1191–1196.

Mayr, E. (1988). *Towards a New Philosophy of Biology: Observations of an Evolutionist.* Harvard University Press, Cambridge.

Mayr, E. (1997). *This is Biology.* Harvard University Press, Cambridge.

Monod, J. (1971). *Chance and Necessity.* Fontana Books. Glasgow.

Morange M. (2012). The recent evolution of the question "What is life?". *History and Philosophy of the Life Sciences* 34, 425–438.

Muller, H.J. (1922). Variations due to change in the individual gene. *The American Naturalist*, 56, 32–50.

Muller, H.J. (1926) The gene as the basis of life. *Proceedings of the 1st International Congress of Plant Sciences*, Ithaca, 897–921.

Muller H.J. (1960). "The Origin of Life". In *Evolution after Darwin: The University of Chicago Centennial Discussions*, Tax, S. and Callender, C. (eds). The University of Chicago Press, Chicago.

Muller, H.J. (1961). Genetic nucleic acid: Key material in the origin of life. *Perspectives in Biology and Medicine*, 5, 1–23.

Peretó, J., Bada, J.L., Lazcano, A. (2009). Charles Darwin and the origins of life. *Origins of Life and Evolution of Biospheres*, 39, 395–406.

Popper, K. (1974). Reduction and the incompleteness of science. In *Studies in the Philosophy of Biology*, Ayala, F. and Dobzhanzky, T. (eds). University of California Press, Berkeley.

Summers W.C. (1999). *Felix d'Herelle and the Origins of Molecular Biology.* Yale University Press, New Haven.

16

Chance, Complexity and the Idea of a Universal Ethics[1]

If all physical interaction is a sort of computation in opposition to chance, then there is a need to rethink cosmic and biological evolution and perhaps even ethics. The universe seems to be gradually organizing itself. We are speaking of "complexification" (Reeves 1981, 1987, 2013; Casti 1994; Delahaye and Vidal 2018). Matter and energy are of course essential to understanding this long-term evolution of the cosmos (Chaisson 2011); however, it is also possible that overemphasizing this prevents us from truly grasping what is happening.

Information and computation theories are a mathematical field for understanding computers, talking about their power, measuring and comparing algorithms, and giving a precise meaning to the word *complexity*. They offer a vision of what the evolution of the universe is in non-material and non-energetic terms. Although considered to be abstract, what they explain seems to account for what we observe in a fairly satisfactory manner, and there is little doubt today that the concepts of *information* and *computation* must be used to understand the dynamics of the physical universe when contemplating it at a large scale.

Researchers who have contributed to this informational and computational vision of the universe are numerous and, without necessarily being in agreement on all points, together they are gradually constructing a new interpretation of cosmic evolution as a computational progression that is opposed to chance, some of which has even been used to deduce ethical considerations. Among the most

Chapter written by Jean-Paul DELAHAYE.

1 This English text is a modified, supplemented and updated version of a French article published in the journal *Pour la science,* no 491, September 2018, pp. 82–87, under the title "L'univers et la morale vus par la théorie du calcul".

important of these researchers, we must cite Andrei Kolmogorov (Li and Vitányi 2008), Leonid Levin (Li and Vitányi 2008), Gregory Chaitin (1990), Charles Bennett (1988), John Mayfield (2013), Luciano Floridi (2010), Seth Lloyd (2006), and more recently, Hector Zenil (Zenil *et al.* 2012), Cédric Gaucherel (2014) and Clément Vidal with whom I have had the pleasure of working with (Delahaye 1999); (Delahaye and Vidal 2018, 2019).

16.1. Cosmic evolution and advances in computation

In understanding cosmic history, astrophysicists have described it as a kind of gradual cooling, revealing more and more complex stable components which interact and combine according to precise rules, leading to atoms and organic molecules, resulting in life. Hubert Reeves (Reeves 1981), presents this now classic vision of a *complexification* of our universe, slow and at a large scale.

Computational theorists interpret this in an abstract form: little by little the universe has given birth to structures that, through interaction, have led to all kinds of computation; due to the stability of these structures, the computation results are sometimes preserved and accumulated. In the case of biological evolution, this led to the development of living beings of increasingly varied and complex forms. Computations with genetic memorization are carried out with increasing efficiency and give rise to new organisms with increasingly elaborate hierarchical structures, which we will call *organized complexity*. In particular, the higher vertebrates have brains to process the data taken in by the senses and retain memories of it. Among humans, language and writing increases the efficiency of information processing and the cumulative storage of results progressively acquired. Information technologies are in turn leading to new advances in efficiency in both computation and the mass storage of information. Over the past 60 years, the capacity of computing per second and the capacity for storing digital information for a given amount have been multiplied by a factor of more than one million. The evolution of the universe is an advance in the computation and cumulative storage of information at the same time as an increase in organized complexity.

It should be noted that this evolution is certainly not monotonous, and that if, observed from afar, the *complexification* is obvious, local regressions, erasures and destruction are frequent. They can cancel out the advances of a period and slow down this powerful but irregular movement that enriches the universe through new forms with improved internal organizations, who interact with their environment in increasingly numerous ways. We should not see any finalism in the vision of this rise in complexity, which is simply a fact that we must come to terms with.

16.2. Two notions of complexity

Andrei Kolmogorov, Leonid Levin and Gregory Chaitin proposed a measure for the complexity of digital objects (texts, images, sounds, videos, etc.), which can be applied to all objects in the universe if we agree to discretize them (Delahaye 1999; Li and Vitányi 2008). This measure is the size of the shortest program generating the digital object Ob, and is called the Kolmogorov complexity of Ob, $K(Ob)$. To memorize a digital object, the best one can do is to reserve a memory of $K(Ob)$ bits of information by placing this shortest computer program that is the optimal compressed version of Ob. The number $K(Ob)$ is the *information content* of the Ob object.

Random digital objects concentrate the largest information content, and therefore have, for a given size in number of bits, the greatest Kolmogorov complexity. Today, the idea is adopted in mathematics to define an infinite random sequence (Martin-Löf 1966). To memorize a random digital object, nothing is significantly better than storing its explicit description without modifying anything. On the other hand, for structured objects, for example the photo of a city or a microprocessor, a list of up to one million prime numbers can be compressed: their information content (Kolmogorov complexity) is significantly smaller than their size. It is wrong to believe that Kolmogorov complexity measures the structural richness of an object. To achieve this, a second measure of complexity must be introduced.

Charles Bennett (Bennett 1988; Delahaye 1999; Li 2008) introduced this measure of organized complexity in 1988 to assess the organizational richness and computational content of digital objects. This is about *logical depth*. As a first approximation, it is the time it takes, measured by the number of computational steps, when starting from the optimal compressed representation of Ob – whose size is $K(Ob)$ – the Ob object is reconstructed. This decompression time measures the richness of Ob structures, since the more structured an object is, the more it offers a subtle means of compressing it, which means that when unwound in reverse, the decompression requires many computational steps. The richness in structures or "organized complexity" is therefore ultimately computational content, which is necessary to convert the optimal compressed Ob form to its explicit form.

Cosmic evolution, by creating organized complexity, therefore accumulates computational content, which is not surprising given that this evolution is also its capacity to store information. Cosmic evolution is a process of perfecting tools that quickly perform computations and preserve the results of those computations, which are structures that are more and more subtle in their organization. In short, cosmic evolution increases – at an irregular and not necessarily monotonous rhythm – the logical depth of the universe and makes it increasingly apt to increase rapidly.

This interpretation appears to be consistent with what is known about the early stages of the universe, the formation of stars and galaxies, and later still the emergence and development of life on earth. The first single-celled beings, higher vertebrates, humans, their cultures and their technologies are stages of progress, both in terms of the organizational richness of the world and in the capacity of worldly objects to efficiently produce new structures and preserve them.

The universe, in a decentralized way of course, becomes a device for computation and memorization, a sort of gigantic computer improving without limitation, with everything everywhere relentlessly producing and preserving objects of increasing organized complexity. The universe computes and stores (partially) the results of its computations; even if it is irregular and sporadic, it increases its logical depth and its ability to produce ever faster.

16.3. Biological computations

The idea that biological evolution should be seen as a computation is championed in particular by John Mayfield (2013). Even though these calculations of evolution and natural selection are quite different from those that a computer performs, it is well known that DNA and RNAs carry information, combine it, copy it and modify and perfect its contents over the long term, and this is the key to modern biology. The mechanisms that cause the appearance of increasingly complex organisms through competition and co-evolution follow a non-monotonous rhythm, where backward steps – possible and frequent – are compensated for and overtaken by steps taken forward in the long term. The relevance of Bennett's concept of *logical depth* in biology is also defended by Antoine Danchin and Cédric Gaucherel (2014), who conducted experiments highlighting its increase in the evolutionary dynamics of simplified ecosystems.

It is important to note that the complexity of Kolmogorov, although constituting a key concept for thinking about the world in informational terms, is not directly usable to talk about the complexification of the universe, as mentioned by astrophysicists and biologists. It is not the information content that increases over time (this content is already at maximum for random objects!) but the content in structures, which is a computational content whose measure is given by Bennett's concept of *logical depth*.

16.4. Energy and emergy

A parallel can be drawn between the concept of computational content measured by Bennett's logical depth and the concept of "emergy" (with an "m") developed and defended by Howard Odum (2007). Emergy for a living, social or technological system is the amount of energy that has been required to create it. Often, this accumulated energy is reduced to solar energy in order to facilitate comparison. Bennett's logical depth is likewise a measure of the amount of computation that has to be done to obtain the object of interest from what is considered to be its probable origin. The two concepts could come together and conversions of evaluations of one into evaluations of the other would be possible... if the energy cost of the computation was constant. Unfortunately, this is not what we are seeing at all. The energy cost of computation, even if we limit ourselves to computations made by our computers, has fallen over the past 60 years by a factor of one million, as mentioned above. More generally, depending on where in the universe we take into account, the energy expended for a computation changes considerably. In addition, a great deal of energy can be expended with nothing left over, if no memory device is present to protect and store the results of computations. The water molecules in the sea interact with each other, which can be seen as a vast parallel computation, however, for the most part, these computations are fading away from moment to moment and the energy that corresponds to this agitation leaves little to no trace.

Bennett's logical depth is a kind of measure of the computational work required to obtain a structure, but this measure only takes into account the computations whose results persist, directly or indirectly. The computational content that measures organized complexity cannot be evaluated in terms of energy: that is something else entirely!

Based on the concepts of information theory and computation, a classification of objects in the world can be proposed. Basically, there are four types of objects in the universe:

1) simple objects (a crystal, a virgin computer memory, etc.). These are the objects with low Kolmogorov complexity and low Bennett logical depth;

2) complex unstructured or very unstructured objects (a pile of sand, a gas, the result of a series of fair dice rolls, etc.). These are the objects with maximum Kolmogorov complexity, which correspond to pure (or almost pure) randomness, but have no computational content and therefore have a low Bennett logical depth;

3) objects resulting from prolonged pure computation that keep track of this, such as the first billion decimal places of the number π, or the drawing of a

microprocessor. Their Kolmogorov complexity is low in contrast to their fairly significant Bennett logical depth;

4) objects associating randomness and richness in structures, such as a living being, a city, an advanced technological object, etc. Both their measures of complexity are high.

16.5. What we hold onto

Armed with this concept of computational content and thinking about the things we value, we see a form of consistency and unity that we love and defend most often.

Human life is, of course, something we want to protect. Each human life is a richness of accumulated experiences, in culture, in intelligence, and also in its capacity to produce its own complex objects. Art and literature create objects which, either directly or through the position they occupy in human societies, enrich the information they produce, as well as their structural complexity. Music, in particular, is a game of varied sound constructions; each new composition giving rise to organizations of rhythms, harmonies and melodies which complement the repertoires already known and preserved.

The sciences, by enriching the relationship between the world and the image we have of it, as well as our means of action, help build new structures in the universe. These structures are perfected through research, that is to say by the organization of experiments (comparable to computations) and the progress of physical and mathematical theories (also comparable to computations), which allow us to formulate hypotheses and select them (all of this is still computation) in order to retain and memorize only the best.

16.6. Noah knew this already!

The majority believe that preserving endangered species is a moral necessity and that letting a species go extinct is far more serious than killing an animal of a species within a thriving population. This moral judgment fits perfectly with what is suggested by Bennett's *principle of maximizing logical depth*. This logical depth of the entire biosphere changes very little when an animal of a large species disappears, while it decreases more markedly when the last representative of a species dies. Preserving animal diversity means preserving organized complexity on earth, preserving its logical depth. In the Bible, when Noah filled his ark, he tried to conserve the diversity of the animal world. With limited means at his disposal, he only took a small number of pairs of each species. The claim that we have a duty to keep the complexity of the world as great as possible is not new: the writers of the Bible acknowledged this three millennia ago! The more general principles of ecological morality stating that we must preserve our environment are cut from the same cloth: the richness of the interactions of animal and plant species in an ecosystem is an organization that was established slowly and which we have a duty to preserve, because once it has been destroyed we cannot reconstitute it.

Destroying a book of which there is only one copy, burning paintings which have no copies and sacking or demolishing monuments, sculptures, ancient objects that bear witness to vanished civilizations all seems ethically incorrect to us. Anything that helps bring down the organized *complexity* of the world is considered to be a bad thing by humans.

16.7. Create, protect and collect

Conversely, creating, preserving and collecting are considered good things for the majority of humans. We are prepared to spend money and resources to ensure that ancient works of art are collected, restored, carefully cataloged, photographed and stored. Our museums are the simplest proof that we love organized complexity and that we make a point of preserving it in all of its diversity.

In us, it seems that, without formulating it explicitly, we seek to increase the organized complexity of the world. A moral principle on which many humans agree requires that we protect, and that we give ourselves the means to always increase this wealth of organized structures in the world. This includes living things, humans in particular, but also works of art, scientific theories, social organizations and everything that sustains and facilitates the flourishing and multiplication of these things.

16.8. An ethics of organized complexity

The result of this analysis is that it is conceivable to formulate a universal ethic of organized complexity: the good would be to conserve, create and contribute to the growth of organized complexity, that is to say the computational contents of the universe measured by Bennett's logical depth.

Charles Bennett, Clément Vidal and I are trying to clarify the ethics that could serve as common ground and the meeting point for a large number of moralities. The idea of ethics being linked to information had already been defended in a different form in 2003 by the philosopher Luciano Floridi (2003, 2010), who wanted to distinguish useful information, which had value for him and which we must consider as a good and useless information like that contained in a series of random draws. Unfortunately, without relying on the concept of logical depth, he could not give his intuitions the rigorous character that is now possible because of the mathematical advances of Charles Bennett.

No one can demonstrate an ethic, therefore the ethic of organized complexity will not claim to be demonstrated. Its defenders can only note several points in its favor. On one hand, it is already present in a large majority of moral and ethical systems, which meet and converge on the imperatives it formulates; it would therefore be a way of defining common ground, bringing together a large number of human beings without contradicting what they already adopt as values, at least in part.

A second element in favor of ethics of organized complexity is that it promotes principles that can be qualified as universal in two complementary ways:

1) it is opposed to nationalisms, speciesism (beliefs that certain species deserve consideration to the exclusion of all others) and to narrow morals that only value very particular beings or objects. It promotes a generous and broad idea of what should be preserved and encouraged;

2) it can be adopted by all autonomous beings with the ability to make decisions: whether human, animal, extraterrestrial (if any) and even robot.

At a time when we are beginning to envision that we might have to live with intelligent beings (robots) with a nature different from our own, it is interesting to know that a simply formulated ethic that is not only anthropocentric is possible. The science fiction idea that we may also have to get along with life on other worlds, and that we will therefore have to share values with them and think about good and bad, without only referring to human beings or to the living world here, also makes this ethic interesting. The question has already been asked (Randolph and McKay 2014; Smith 2014) and all possible answers point in the direction of broadening recent moral considerations, which already suggested that we must preserve the world around us and stop degrading it under the pretence that only humanity matters. An ethic of complexity conforms to these expanded visions of values, while at the same time confirming that the arts and sciences create forms of wealth that we have a duty to encourage and protect.

16.9. Not so easy

Although the general idea is simple and rests on a precise mathematical concept, not everything would become easy for someone who adopted ethics of complexity based on Bennett's logical depth. Those who adopt this ethic of organized complexity would have to face at least three obstacles:

– organized complexity is not easy to assess. Even if our intuition often allows us to know what holds this complexity and what does not, this is not always the case, and in any case the comparisons remain delicate: who can claim to judge with certainty that certain music contains a richness in structures greater than another? Even if a mathematical definition exists, it is not easy to derive precise, or even approximate, measurement tools from it, as any attempt at measurement encounters the obstacle of the combinatorial explosion of computations that need to be carried out: measuring computational content requires a lot of computation;

– it is not clear whether we should seek to maximize the organized complexity of the world, or the average organized complexity of the world, which does not amount to the same thing: on one hand, we will avoid any copying, and on the other hand, a great deal of importance will be given to the circulation, the diffusion and the multiplication of copies of a complex structure. There are other similar problems;

– faced with a choice between several actions, determining which one will best satisfy the principle of maximizing organized complexity will not be obvious since it will be necessary to anticipate the effects of the choice, and therefore formulate bets on the triggered causal chains of the various options.

One needs to be careful, however, to ensure that these challenges do not make it impossible or absurd to adopt ethics of organized complexity. The three problems mentioned also exist with regard to all other ethics. Imagine, for example, that we consider that good is everything that contributes to human happiness, an idea quite often accepted implicitly in politics and acceptable to many minds. Taking such a humanist ethic of happiness seriously and wanting to apply it rigorously would require having a clear definition of what human happiness is and the tools to measure it. This is obviously not the case: there are no unanimous methods of evaluating human happiness, since each individual has their own conception of this.

Whoever takes human happiness as a goal, even if they knew how to evaluate it, would also have to choose between maximizing the average happiness of humans, the happiness of the most unhappy, or the happiness of the happiest (an unlikely option, and yet we wonder whether this is the direction our societies are heading toward). These questions, even more complicated than those linked to the sharing of wealth, are not unanimous and are, on the contrary, the subject of divergent political options.

Faced with each decision to be made, it also remains to predict what will result in the medium term and long term from the various options and their consequences for the type of human happiness that we have chosen to promote. This is the problem of anticipating effects.

We can see that the three mentioned difficulties encountered by an ethics of organized complexity have their exact counterparts for ethics of human happiness. The objections to be made to ethics of organized complexity therefore cannot relate to the difficulty there is in implementing it, which is no greater than the difficulty encountered by any ethic that attempts to formulate itself with precision.

An ethic is not a fact, true or false, it is only a conception that one adopts (or not) in order to determine what is right and wrong. The reflections one finds oneself engaged in when considering ethics of organized complexity certainly deserves profound inspection, especially since this would lead to an elegant morality and aims that allow for more self-transcendence than strictly humanist ones, or those centered on narrow values, or lists of rules and commandments to be observed. A new challenge is presented to all: to design scientifically founded moral ideals, as universal as possible, for us and our descendants, which value humans, life in general, the arts and sciences and which are adapted toward an expanding world of robots, and perhaps life on other planets (Zenil 2012; Bennett 2014; Delahaye and Vidal 2018; Delahaye 2019).

16.10. References

Bennett, C. (1988). Logical depth and physical complexity. In *The Universal Turing Machine: A Half-Century Survey*, Herken, R. (ed.). Oxford University Press [Online]. Available at: https://pdfs.semanticscholar.org/ac97/5f088cf61c09bae8506808468a08467d 55e6.pdf.

Bennett, C. (2014). Evidence, computation, and ethics. Simons Symposium on Evidence in the Natural Sciences [Online]. Available at: https://simonsfoundation.s3.amazonaws.com/ share/mps/conferences/Symposium_on_Evidence_in_the_Natural_Sciences/Bennett_slides.pdf.

Casti, J. (1994). *Complexification: Explaining a Paradoxical World through the Science of Surprise*. Springer, New York.

Chaisson, E. (2011). Energy rate density as a complexity metric and evolutionary driver. *Complexity*, 16(3), 27–40.

Chaitin, G. (1990). *Information, Randomness & Incompleteness: Papers on Algorithmic Information Theory*. World Scientific, Singapore.

Delahaye, J.-P. (1999). *Information complexité et hasard*. Editions Hermès, Paris.

Delahaye, J.-P. and Vidal, C. (2018). Organized complexity: Is big history a big computation? *American Philosophical Association Newletter on Philosophy and Computer*, 17(2), 49–54 [Online]. Available at: http://c.ymcdn.com/sites/www.apaonline.org/resource/ collection/EADE8D52-8D02-4136-9A2A-729368501E43/ComputersV17n2.pdf.

Delahaye, J.-P. and Vidal, C. (2019). Universal ethics: Organized complexity as an intrinsic value. In *Evolution, Development and Complexity: Multiscale Evolutionary Models of Complex Adaptive Systems*, Georgiev, G.Y., Martinez, C.F., Price, M.E., Smart, J.M. (eds). Springer [Online]. Available at: doi:10.5281/zenodo.1172976 and https://doi.org/10.5281/zenodo.1172976.

Floridi, L. (2003). Two approaches to the philosophy of information. *Minds and Machines*, 13(4), 459–469.

Floridi, L. (ed.) (2010). *The Cambridge Handbook of Information and Computer Ethics*. Cambridge University Press, Cambrige.

Gaucherel, C. (2014). Ecosystem complexity through the lens of logical depth: Capturing ecosystem individuality. *Biological Theory*, 9(4), 440–51.

Li, M. and Vitányi, P. (2008). *An Introduction to Kolmogorov Complexity and Its Applications*. Springer, New York.

Lloyd, S. (2006). *Programming the Universe: A Quantum Computer Scientist Takes on the Cosmos*. Alfred A. Knopf, Random House Inc., New York.

Martin-Löf, P. (1966). The definition of random sequences. *Information and Control*, 9(6), 602–619.

Mayfield, J. (2013). *The Engine of Complexity: Evolution as Computation*. Columbia University Press, New York.

Odum, H. (2007). *Environment, Power, and Society for the Twenty-First Century: The Hierarchy of Energy*. Columbia University Press, New York.

Randolph, R. and McKay, C. (2014). Protecting and expanding the richness and diversity of life, an ethic for astrobiology research and space exploration. *International Journal of Astrobiology*, 13(1), 28–34.

Reeves, H. (1981). *Patience dans l'azur, l'évolution cosmique*. Le Seuil, Paris.

Reeves, H. (1987). *De l'univers à l'homme : l'aventure de la complexification, les scientifiques parlent*. Albert Jacquard (ed.). Hachette, Paris [Online]. Available at: https://www.revue3emillenaire.com/blog/de-lunivers-a-lhomme-laventure-de-la-complexification-par-hubert-reeves/.

Smith, K. (2014). Manifest complexity: A foundational ethic for astrobiology? *Space Policy*, 30(4), 209–214.

Zenil, H., Delahaye, J.-P., Gaucherel, C. (2012). Image characterization and classification by physical complexity. *Complexity*, 17(3), 26–42.

List of Authors

Georges AMAR
Chair of Design Theory and
Methods for Innovation
MINES ParisTech
France

Cristian S. CALUDE
School of Computer Science
University of Auckland
New Zealand

Michel CASSÉ
CEA
Paris
France

Jean-Paul DELAHAYE
University of Lille
CRIStAL
UMR-CNRS
Villeneuve d'Ascq
France

Stéphane DOUADY
UMR-CNRS
Paris Diderot University
France

Bernard DUJON
Pasteur Institute
Sorbonne University
Paris
France

Thierry GAUDIN
MINES ParisTech
France

Philippe GRANDCOLAS
ISYEB
National Museum of Natural History
CNRS
Sorbonne University
EPHE, UA
Paris
France

Clarisse HERRENSCHMIDT
CNRS
Paris
France

Amaury LAMBERT
Laboratory of Probability,
Statistics and Modeling
UMR-CNRS
Sorbonne University
and
Centre for Interdisciplinary
Research in Biology
Collège de France,
CNRS, INSERM, PSL
Paris
France

Antonio LAZCANO
El Colegio Nacional
Faculty of Sciences
UNAM
Mexico

Giuseppe LONGO
Centre Cavaillès – République
des Savoirs
CNRS and École Normale Supérieure
Paris
France
and
Tufts University School of Medicine
Boston
USA

Ivan MAGRIN-CHAGNOLLEAU
Aix-Marseille University
PRISM, CNRS
Marseille
France
and
Chapman University
California
USA

Marie-Christine MAUREL
Sorbonne University
and
Institute of Systematics, Evolution,
Biodiversity
MNHN
Paris
France

Gilles PAGÈS
Laboratory of Probability,
Statistics and Modeling
UMR-CNRS
Sorbonne University
France

Mathias PESSIGLIONE
Paris Brain Institute
Pitié-Salpêtrière Hospital
France

Jean-Charles POMEROL
Professor Emeritus
Sorbonne University
Paris
France

David SITBON
Institut Curie
PSL Research University
Sorbonne University
UMR-CNRS
Equipe Labellisée Ligue Contre le
Cancer
Paris
France

François VANNUCCI
APC/LPNHE
Paris Diderot University
France

Bertrand VERGELY
Professor of Philosophy
Lycée Pothier
Orléans
and
Sciences Po
Paris
and
Institut de théologie Saint Serge
Paris
France

Jonathan B. WEITZMAN
The Centre for Epigenetics
and Cell Fate
Paris Diderot University
CNRS
France

Index

Other titles from

in

Information Systems, Web and Pervasive Computing

2020

CLIQUET Gérard, with the collaboration of BARAY Jérôme
Location-Based Marketing: Geomarketing and Geolocation

DE FRÉMINVILLE Marie
Cybersecurity and Decision Makers: Data Security and Digital Trust

EL ASSAD Safwan, BARBA Dominique
Digital Communications 1: Fundamentals and Techniques

EL ASSAD Safwan, BARBA Dominique
Digital Communications 2: Directed and Practical Work

GEORGE Éric
Digitalization of Society and Socio-political Issues 2: Digital, Information and Research

HELALI Saida
Systems and Network Infrastructure Integration: Design, Implementation, Safety and Supervision

LOISEAU Hugo, VENTRE Daniel, ADEN Hartmut
Cybersecurity in Humanities and Social Sciences: A Research Methods Approach (Cybersecurity Set – Volume 1)

SEDKAOUI Soraya, KHELFAOUI Mounia
Sharing Economy and Big Data Analytics

SCHMITT Eglantine
Big Data (Intellectual Technologies Set – Volume 1)

2019

ALBAN Daniel, EYNAUD Philippe, MALAURENT Julien, RICHET Jean-Loup, VITARI Claudio
Information Systems Management: Governance, Urbanization and Alignment

AUGEY Dominique, with the collaboration of ALCARAZ Marina
Digital Information Ecosystems: Smart Press

BATTON-HUBERT Mireille, DESJARDIN Eric, PINET François
Geographic Data Imperfection 1: From Theory to Applications

BRIQUET-DUHAZÉ Sophie, TURCOTTE Catherine
From Reading-Writing Research to Practice

BROCHARD Luigi, KAMATH Vinod, CORBALAN Julita, HOLLAND Scott, MITTELBACH Walter, OTT Michael
Energy-Efficient Computing and Data Centers

CHAMOUX Jean-Pierre
The Digital Era 2: Political Economy Revisited

COCHARD Gérard-Michel
Introduction to Stochastic Processes and Simulation

DUONG Véronique
SEO Management: Methods and Techniques to Achieve Success

GAUCHEREL Cédric, GOUYON Pierre-Henri, DESSALLES Jean-Louis
Information, The Hidden Side of Life

GEORGE Éric
Digitalization of Society and Socio-political Issues 1: Digital, Communication and Culture

GHLALA Riadh
Analytic SQL in SQL Server 2014/2016

JANIER Mathilde, SAINT-DIZIER Patrick
Argument Mining: Linguistic Foundations

SOURIS Marc
Epidemiology and Geography: Principles, Methods and Tools of Spatial Analysis

TOUNSI Wiem
Cyber-Vigilance and Digital Trust: Cyber Security in the Era of Cloud Computing and IoT

2018

ARDUIN Pierre-Emmanuel
Insider Threats
(Advances in Information Systems Set – Volume 10)

CARMÈS Maryse
Digital Organizations Manufacturing: Scripts, Performativity and Semiopolitics
(Intellectual Technologies Set – Volume 5)

CARRÉ Dominique, VIDAL Geneviève
Hyperconnectivity: Economical, Social and Environmental Challenges
(Computing and Connected Society Set – Volume 3)

CHAMOUX Jean-Pierre
The Digital Era 1: Big Data Stakes

DOUAY Nicolas
Urban Planning in the Digital Age
(Intellectual Technologies Set – Volume 6)

FABRE Renaud, BENSOUSSAN Alain
The Digital Factory for Knowledge: Production and Validation of Scientific Results

GAUDIN Thierry, LACROIX Dominique, MAUREL Marie-Christine, POMEROL Jean-Charles
Life Sciences, Information Sciences

GAYARD Laurent
Darknet: Geopolitics and Uses
(Computing and Connected Society Set – Volume 2)

IAFRATE Fernando
Artificial Intelligence and Big Data: The Birth of a New Intelligence
(Advances in Information Systems Set – Volume 8)

LE DEUFF Olivier
Digital Humanities: History and Development
(Intellectual Technologies Set – Volume 4)

MANDRAN Nadine
Traceable Human Experiment Design Research: Theoretical Model and Practical Guide
(Advances in Information Systems Set – Volume 9)

PIVERT Olivier
NoSQL Data Models: Trends and Challenges

ROCHET Claude
Smart Cities: Reality or Fiction

SALEH Imad, AMMI, Mehdi, SZONIECKY Samuel
Challenges of the Internet of Things: Technology, Use, Ethics
(Digital Tools and Uses Set – Volume 7)

SAUVAGNARGUES Sophie
Decision-making in Crisis Situations: Research and Innovation for Optimal Training

SEDKAOUI Soraya
Data Analytics and Big Data

SZONIECKY Samuel
Ecosystems Knowledge: Modeling and Analysis Method for Information and Communication
(Digital Tools and Uses Set – Volume 6)

2017

BOUHAÏ Nasreddine, SALEH Imad
Internet of Things: Evolutions and Innovations
(Digital Tools and Uses Set – Volume 4)

DUONG Véronique
Baidu SEO: Challenges and Intricacies of Marketing in China

LESAS Anne-Marie, MIRANDA Serge
The Art and Science of NFC Programming
(Intellectual Technologies Set – Volume 3)

LIEM André
Prospective Ergonomics
(Human-Machine Interaction Set – Volume 4)

MARSAULT Xavier
Eco-generative Design for Early Stages of Architecture
(Architecture and Computer Science Set – Volume 1)

REYES-GARCIA Everardo
The Image-Interface: Graphical Supports for Visual Information
(Digital Tools and Uses Set – Volume 3)

REYES-GARCIA Everardo, BOUHAÏ Nasreddine
Designing Interactive Hypermedia Systems
(Digital Tools and Uses Set – Volume 2)

SAÏD Karim, BAHRI KORBI Fadia
Asymmetric Alliances and Information Systems:Issues and Prospects
(Advances in Information Systems Set – Volume 7)

SZONIECKY Samuel, BOUHAÏ Nasreddine
Collective Intelligence and Digital Archives: Towards Knowledge Ecosystems
(Digital Tools and Uses Set – Volume 1)

2016

BEN CHOUIKHA Mona
Organizational Design for Knowledge Management

BERTOLO David
Interactions on Digital Tablets in the Context of 3D Geometry Learning
(Human-Machine Interaction Set – Volume 2)

BOUVARD Patricia, SUZANNE Hervé
Collective Intelligence Development in Business

EL FALLAH SEGHROUCHNI Amal, ISHIKAWA Fuyuki, HÉRAULT Laurent, TOKUDA Hideyuki
Enablers for Smart Cities

FABRE Renaud, in collaboration with MESSERSCHMIDT-MARIET Quentin, HOLVOET Margot
New Challenges for Knowledge

GAUDIELLO Ilaria, ZIBETTI Elisabetta
Learning Robotics, with Robotics, by Robotics
(Human-Machine Interaction Set – Volume 3)

HENROTIN Joseph
The Art of War in the Network Age
(Intellectual Technologies Set – Volume 1)

KITAJIMA Munéo
Memory and Action Selection in Human–Machine Interaction
(Human–Machine Interaction Set – Volume 1)

LAGRAÑA Fernando
E-mail and Behavioral Changes: Uses and Misuses of Electronic Communications

LEIGNEL Jean-Louis, UNGARO Thierry, STAAR Adrien
Digital Transformation
(Advances in Information Systems Set – Volume 6)

NOYER Jean-Max
Transformation of Collective Intelligences
(Intellectual Technologies Set – Volume 2)

VENTRE Daniel
Information Warfare – 2nd edition

VITALIS André
The Uncertain Digital Revolution
(Computing and Connected Society Set – Volume 1)

2015

ARDUIN Pierre-Emmanuel, GRUNDSTEIN Michel, ROSENTHAL-SABROUX Camille
Information and Knowledge System
(Advances in Information Systems Set – Volume 2)

BÉRANGER Jérôme
Medical Information Systems Ethics

BRONNER Gérald
Belief and Misbelief Asymmetry on the Internet

IAFRATE Fernando
From Big Data to Smart Data
(Advances in Information Systems Set – Volume 1)

KRICHEN Saoussen, BEN JOUIDA Sihem
Supply Chain Management and its Applications in Computer Science

NEGRE Elsa
Information and Recommender Systems
(Advances in Information Systems Set – Volume 4)

POMEROL Jean-Charles, EPELBOIN Yves, THOURY Claire
MOOCs

2012

BUCHER Bénédicte, LE BER Florence
Innovative Software Development in GIS

GAUSSIER Eric, YVON François
Textual Information Access

STOCKINGER Peter
Audiovisual Archives: Digital Text and Discourse Analysis

VENTRE Daniel
Cyber Conflict

2011

BANOS Arnaud, THÉVENIN Thomas
Geographical Information and Urban Transport Systems

DAUPHINÉ André
Fractal Geography

LEMBERGER Pirmin, MOREL Mederic
Managing Complexity of Information Systems

STOCKINGER Peter
Introduction to Audiovisual Archives

STOCKINGER Peter
Digital Audiovisual Archives

VENTRE Daniel
Cyberwar and Information Warfare

2010

BONNET Pierre
Enterprise Data Governance

BRUNET Roger
Sustainable Geography

KANEVSKI Michael
Advanced Mapping of Environmental Data

MANOUVRIER Bernard, LAURENT Ménard
Application Integration: EAI, B2B, BPM and SOA

PAPY Fabrice
Digital Libraries

2007

DOBESCH Hartwig, DUMOLARD Pierre, DYRAS Izabela
Spatial Interpolation for Climate Data

SANDERS Lena
Models in Spatial Analysis

2006

CLIQUET Gérard
Geomarketing

CORNIOU Jean-Pierre
Looking Back and Going Forward in IT

DEVILLERS Rodolphe, JEANSOULIN Robert
Fundamentals of Spatial Data Quality

Printed in the USA
CPSIA information can be obtained
at www.ICGtesting.com
CBHW070752290524
9015CB00051B/12